The Physiology of the INSECT CENTRAL NERVOUS SYSTEM

The Physiology of the INSECT CENTRAL NERVOUS SYSTEM

Papers from the 12th International Congress of Entomology held in London, 1964

Edited by

J. E. TREHERNE and J. W. L. BEAMENT
Department of Zoology, University of Cambridge, England

1965

ACADEMIC PRESS
LONDON AND NEW YORK

ACADEMIC PRESS INC. (LONDON) LTD
Berkeley Square House
Berkeley Square
London, W.1

U.S. Edition published by
ACADEMIC PRESS INC.
111 Fifth Avenue
New York, New York, 10003

Library of Congress Catalog Card Number: 65-18446

PRINTED IN GREAT BRITAIN BY
BUTLER AND TANNER LTD
FROME, SOMERSET

List of Contributors

BERNARD, J., *Laboratoire de Physiologie Animale, Faculté des Sciences de Rennes, France.*

BOISTEL, J., *Laboratoire de Physiologie Animale, Faculté des Sciences de Rennes, France.*

CALLEC, J. J., *Laboratoire de Physiologie Animale, Faculté des Sciences de Rennes, France.*

GAHERY, Y., *Laboratoire de Physiologie Animale, Faculté des Sciences de Rennes, France.*

HORRIDGE, G. A., *Gatty Marine Laboratory, The University, St. Andrews, Fife, Scotland.*

HOYLE, G., *Department of Biology, University of Oregon, Eugene, Oregon, U.S.A.*

HUBER, F., *Department of Zoology, University of California, Los Angeles, California, U.S.A.*

HUGHES, G. M., *Department of Zoology, University of Cambridge, Cambridge, England.*

MILLER, P. J., *Department of Zoology, University of Oxford, England.*

NARAHASHI, T., *Laboratory of Applied Entomology, Faculty of Agriculture, University of Tokyo, Japan.*

RAY, J. W., *Agricultural Research Council, Pest Infestation Laboratory, Slough, Buckinghamshire, England.*

ROEDER, K. D., *Department of Biology, Tufts College, Medford, Massachusetts, U.S.A.*

ROWELL, C. H. F., *Department of Zoology, Makerere University College, Kampala, Uganda, East Africa.*

SCHOLES, J. H., *Gatty Marine Laboratory, The University, St. Andrews, Fife, Scotland.*

SHAW, S., *Gatty Marine Laboratory, The University, St. Andrews, Fife, Scotland.*

SMITH, D. S., *Department of Zoology, University of Virginia, Charlottesville, Virginia, U.S.A.*

TREHERNE, J. E., *Department of Zoology, University of Cambridge, Cambridge, England.*

TUNSTALL, J., *Gatty Marine Laboratory, The University, St. Andrews, Fife, Scotland.*

WEEVERS, R. DE G., *Department of Zoology, University of Cambridge, Cambridge, England.*

WILSON, D. M., *Department of Zoology, University of California, Berkeley, California, U.S.A.*

Preface

The relative simplicity and anatomical disposition of the insect central nervous system makes it an ideal medium for the study of its physiology. This volume reports the first major Symposium to be attended by authorities in this field which was held at the 12th International Congress of Entomology in London in July 1964. The contributions are of two kinds: review articles, which are valuable to the general biologist as well as to the specialist, and papers dealing with the most recent discoveries in several aspects of the subject. These contributions represent a continuous spectrum, ranging from investigations on the physiology of giant axons, the chemistry of nervous tissues and the mechanism of synaptic transmission to such topics as the neuronal pathways in the central nervous system, the central control of physiological processes, and finally to the neurophysiological basis of learning and instinctive behaviour in insects.

As Professor Roeder emphasizes in his far-sighted epilogue, our progress in understanding the physiology of the central nervous system depends not only on advances in technique, but also upon our way of thinking analytically about it. It is hoped that the contributions to this volume will serve both as a useful summary of our present state of knowledge and also as a stimulus to further research and analysis in this subject.

J. E. T.

March, 1965 J. W. L. B.

Contents

Contents

The Physiology of Insect Axons

TOSHIO NARAHASHI

Laboratory of Applied Entomology, Faculty of Agriculture,
University of Tokyo, Tokyo, Japan

The neurophysiology of insects relies, in part at least, upon a knowledge of general neurophysiology, although there are certain fields in which work on insects has influenced the general interpretation of nervous functions. The physiology of insect nerve and muscle is important for two reasons: (1) because of the enormous variety of insects (in terms of morphology, physiology or behaviour) which provides us with excellent opportunity to study comparative neurophysiology, and (2) since many of the currently developed insecticides are neurotoxins, it is of great importance to know the basic aspects of insect nervous function in order to elucidate the mode of action of these insecticides.

The study of neurophysiology is based in many cases on the observations of potential changes occurring in nerve and muscle. The technique is of use for either of two purposes: (1) to learn at the cellular or even molecular level how nerve and muscle work, and (2) to learn how organ and animal are organized and work. For instance, the mechanism of action potential production belongs to the former category, while the behaviouristic responses induced by external stimulation belong to the latter. The cellular aspect provides the latter with the basic knowledge.

Neurophysiology has achieved remarkable progress in the recent two decades so that a variety of nervous functions can now be interpreted in physico-chemical terms. The first measurements of the action potentials of nerve and muscle were made using external electrodes, which only permitted observation of severely attenuated potential changes due to short circuit. Subsequently, however, Curtis and Cole (1940) and Hodgkin and Huxley (1939) were able to measure the absolute magnitudes of membrane resting and action potentials by introducing a longitudinal electrode into a squid giant axon. The action potentials recorded in these studies showed an overshoot beyond the zero potential level, a discovery which was not accounted for by the classical membrane theory of Bernstein (1912) which predicted a simple disappearance of the membrane potential during activity. This discovery stimulated physiologists, especially those of the Cambridge group headed by Professor A. L. Hodgkin, to explore the mechanisms causing the overshoot of the action potential. Another technical achievement was the voltage-clamp method which was first developed by Marmont (1949) and Cole (1949) and then extensively used by Hodgkin's and other groups (e.g. Cole and Moore, 1960; Hagiwara and Saito, 1959a, b; Hodgkin, Huxley and Katz, 1949, 1952; Julian, Moore and Goldman, 1962b; Narahashi, Moore and Scott, 1964; Tasaki and Hagiwara, 1957). This technique permits us to analyse the sequence of events occurring

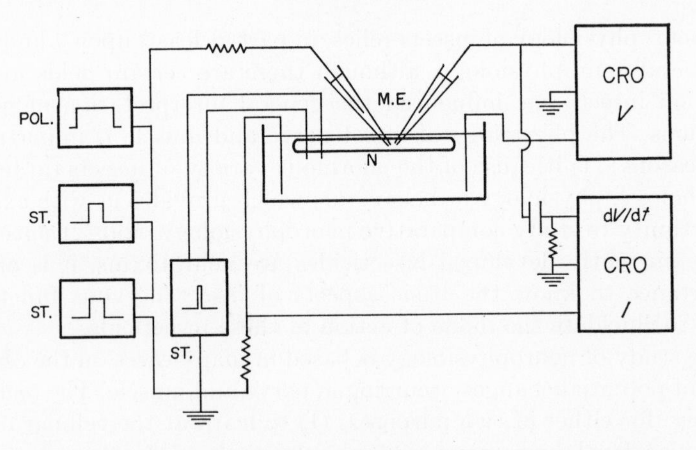

FIG. 1. Diagram of stimulation and recording by means of intracellular microelectrodes. CRO, cathode ray oscilloscope; dV/dt, recording of the rates of rise and fall of the action potential; I, current recording; M.E., microelectrode; N, nerve; POL., polarization; ST., stimulation; V, voltage recording.

at the nerve membrane in terms of membrane current and membrane conductance, which are otherwise too complicated to analyse. An additional tool, the radio-active isotope, has also been of value in the direct measurement of ionic fluxes across the nerve and muscle membrane at rest and during activity.

In the present article, the most fundamental aspects of insect cellular neurophysiology are described, and attempts are made to apply this knowledge to elucidating the mechanism of drug action on insect nerve, and to extend it to the molecular level of excitation mechanisms. Analyses have been made in cockroach giant axons by means of the intracellular microelectrode technique unless otherwise stated (Fig. 1).

I. Physiological Aspects of Nervous Functions

A. *Membrane Electrical Properties*

A nerve fibre is composed of an axoplasm surrounded by a very thin membrane, the "nerve membrane" or "excitable membrane". It is at this membrane that the nerve manifests various physiological functions such as impulse conduction. The nerve membrane, about 100 Å thick, is made of protein and phospholipid molecules. There is a potential difference of 50–100 mV across the nerve membrane, inside being negative with respect to outside; this is called the "resting potential". The nerve membrane shows a relatively high electrical resistance parallel with a capacity (Fig. 2). Since the resistances of the

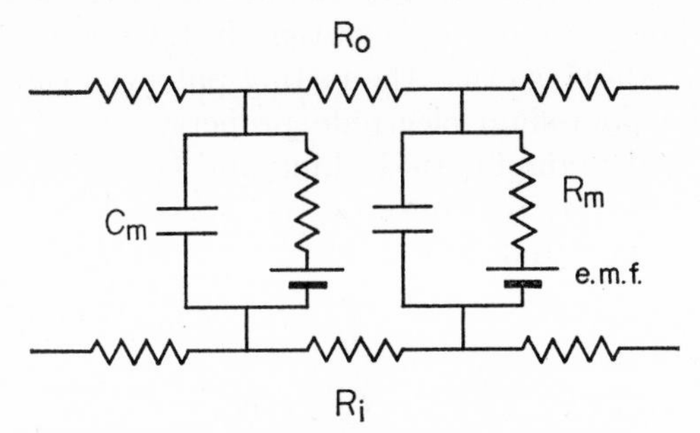

Fig. 2. Electrical equivalent circuit of the nerve membrane. C_m, membrane capacity; R_i, internal resistance; R_m, membrane resistance; R_o, external resistance.

axoplasm and of the external fluid are relatively low, a nerve fibre behaves electrically like a cable. Because of such a cable structure, a potential difference imposed at a point on the nerve fibre falls decrementally with distance along the fibre and also changes with time (Fig. 3). It is possible to estimate the values of the electrical constants

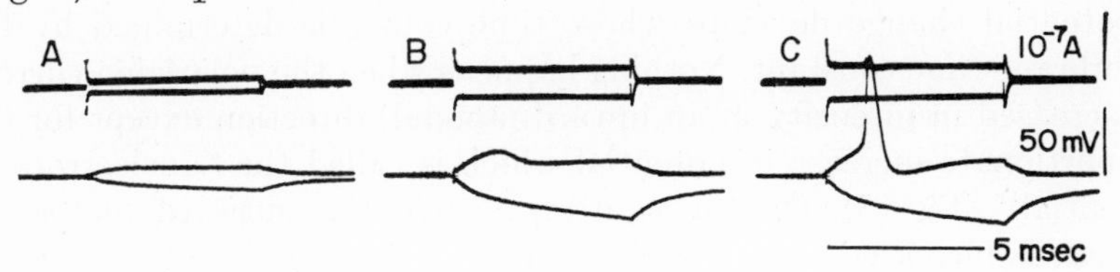

Fig. 3. Potential changes produced by passing square pulses of polarizing current through the membrane of a cockroach giant axon. Upper tracings represent current recording and the zero potential level, upward deflexion being outward current. Lower tracings represent potential recording, upward deflexion being depolarization. Responses to outward and inward currents of the same intensity are superimposed in each set of the records. Note the local response and delayed rectification in B and the action potential superimposed on a catelectrotonic potential in C. Preparation 61–219Ba. Temperature 20° C.

by applying square pulses of current to a nerve fibre and by recording the resultant potential changes. Table I gives results of such experiments, with some examples from other animals for comparison. It is apparent that the electrical properties of the cockroach giant axon in large measure resemble those of other nerves.

B. *Nature of Resting Potential*

The resting potential of the cockroach giant axons is of the same order of magnitude as that of other nerves (Table I). It is reduced by increasing the external concentration of potassium (Fig. 2 of Yamasaki and Narahashi, 1959a). As in other nerves and muscles, the potassium concentration is higher in the axoplasm than in the external fluid (Tobias, 1948; Treherne, 1961), so that the resting potential can be interpreted in terms of the potassium electrode property (e.g. Curtis and Cole, 1942; Huxley and Stämpfli, 1951; Ling and Gerard, 1950); that is to say, it is determined by the concentration gradient of potassium across the nerve membrane, and approaches the potassium equilibrium potential, E_K,

$$E_K = \frac{RT}{F} \ln \frac{[K]_o}{[K]_i}, \qquad (1)$$

where R is the gas constant, T is the absolute temperature, F is the Faraday constant, and $[K]_o$ and $[K]_i$ are potassium concentrations outside and inside the nerve, respectively.

C. *Nature of Action Potential*

When a cockroach giant axon is subjected at one point to a weak pulse of current of either potentiality with respect to the membrane, a potential change develops whose time course is determined by the membrane time constant. Nothing happens when the polarizing current is increased in intensity in an inward (anodal) direction except for the proportionate increase in potential which is called the "anelectrotonic potential" (Fig. 3). On the contrary, when the outward (cathodal) current is increased in intensity, there is a critical point of depolarization beyond which the depolarization grows rapidly, crosses the zero potential level and then returns towards the resting potential level. These changes are known as the "action potential" (Figs. 3 and 6A). When the action potential occurs at one point of the axon the neighbouring region is subjected to outward membrane current, due to the potential difference established, so as to produce another action potential (Fig. 4). Because of the regenerative nature of the action potential production, the impulse is conducted along the axon without undergoing any decrement. When the properties of the axon are changed in such a way as to lower the safety factor of conduction, impulse conduction is blocked.

The action potential of the cockroach giant axon is of the same order of magnitude as that of other nerves (Table I). It is reduced by decreasing the external sodium concentration (Fig. 6 of Yamasaki and Narahashi, 1959a), the relation being again very similar to that found in other nerves and muscles (e.g. Hodgkin and Katz, 1949; Huxley and

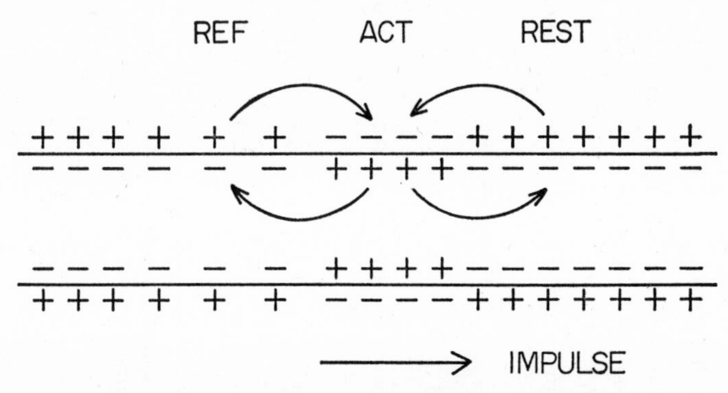

FIG. 4. Diagram of impulse conduction along a nerve fibre. ACT, active region; REF, refractory region; REST, resting region.

Stämpfli 1951; Nastuk and Hodgkin, 1950). This is to be expected when the action potential behaves as a sodium electrode, i.e.

$$E_{Na} = \frac{RT}{F} \ln \frac{[Na]_o}{[Na]_i}, \tag{2}$$

where E_{Na} is the sodium equilibrium potential, and $[Na]_o$ and $[Na]_i$ are sodium concentrations outside and inside the nerve, respectively. Unlike potassium, the sodium concentration is higher in the external medium than in the axoplasm (Tobias, 1948; Treherne, 1961).

It is then reasonable to assume that the membrane resting and action potentials are explicable in terms of the ionic theory advanced by Hodgkin and his associates (Hodgkin, 1951, 1958). Figure 5 shows the schematic explanation of excitation. In the resting state, the membrane is permeable to potassium but only scarcely so to sodium, so that the membrane potential approaches the potassium equilibrium potential. When the nerve is stimulated, the sodium conductance rises quickly,

TABLE I

MEMBRANE POTENTIALS AND MEMBRANE ELECTRICAL CONSTANTS

Tissue	RP (mV)	AP (mV)	R_m (Ωcm^2)	R_i (Ωcm)	C_m ($\mu F/cm^2$)	τ_m (msec)	λ (mm)	Reference
Cockroach axon	77	99*	800		6·3	4·2	0·86	Narahashi and Yamasaki (1960a), Yamasaki and Narahashi (1959b)
	78	85†	610	46			1·3	Boistel and Coraboeuf (1954), Boistel (1959)
Squid axon	61	104	700	30	1·0	0·7	6·0	Cole and Hodgkin (1939), Curtis and Cole (1942), Hodgkin, Huxley and Katz (1949)
Crab axon	71–94	116–153	7700	90	1·1	6·8	2·0	Hodgkin (1947), Hodgkin and Huxley (1945)
Lobster axon	73	101	2300	60	1·3	2·3	1·6	Hodgkin and Rushton (1946), Tobias and Bryant (1955)

RP, resting potential; AP, action potential; R_m, specific membrane resistance; R_i, specific axoplasm resistance; C_m, specific membrane capacity; τ_m, membrane time constant; λ, membrane length constant.

* In 214 mM-Na Ringer. † In 154 mM-Na Ringer.

thereby causing the membrane potential to approach the sodium equilibrium potential; this is the rising phase of the action potential. At this time sodium enters the axon according to its concentration gradient. The sodium conductance then begins to decrease and the

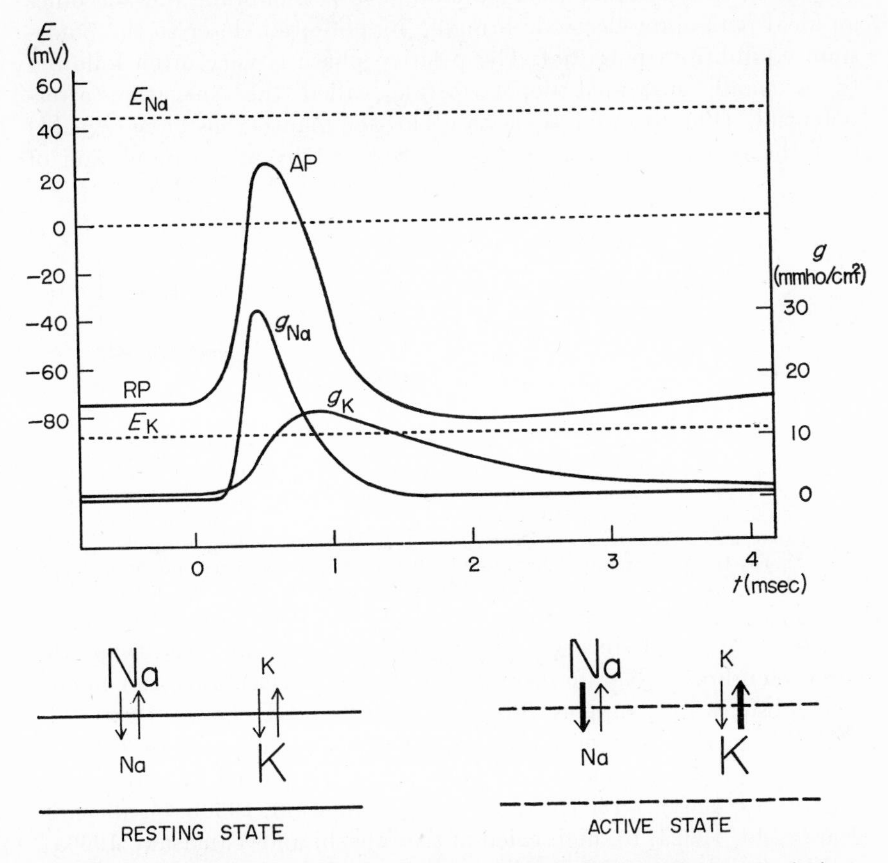

FIG. 5. Schematic representation of the excitation mechanism by the ionic theory. AP, action potential; E, membrane potential; E_K, potassium equilibrium potential; E_{Na}, sodium equilibrium potential; g, membrane conductance; g_K, membrane potassium conductance; g_{Na}, membrane sodium conductance; t, time. The upper scheme is adapted from Hodgkin (1958).

potassium conductance now begins to increase, both working to bring the membrane potential down to the resting level; this is the falling phase of the action potential. Potassium now tends to escape from the axon according to its concentration gradient. In both cockroach and squid giant axons, the falling phase may be followed by an undershoot, which is called the "positive phase" (Fig. 6); this is not seen in lobster giant axons. The positive phase is attributable to the sustained increase in potassium conductance. Because the actual resting potential is slightly lower than the potassium equilibrium potential due to conductances to ions other than potassium, the membrane now becomes an ideal potassium electrode bringing its potential closer to the potassium equilibrium potential. The positive phase is very often followed by a small, sustained depolarization called the "negative after-potential" (Fig. 6). Analyses with cockroach giant axons have revealed that the negative after-potential is produced by an accumulation of

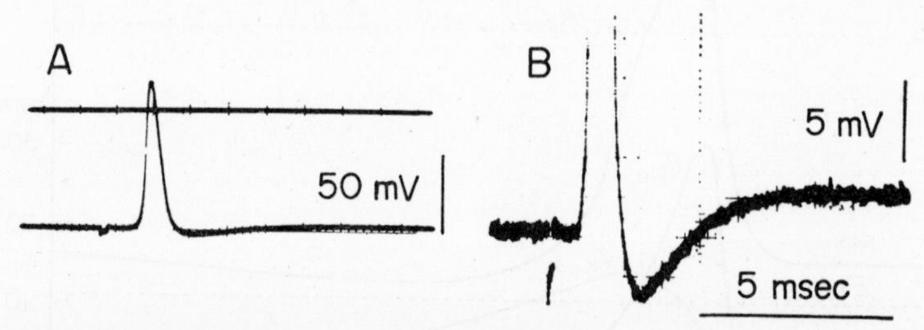

FIG. 6. Propagated action potentials produced by a brief pulse of current in a cockroach giant axon. The spike potential is followed by a positive phase which is terminated in a prolonged negative after-potential. The latter two phases are more clearly seen in B where the voltage amplification is 10 times greater. The upper tracing in A shows the zero potential level. Preparation 64–219Ba. Temperature 20° C.

potassium released during activity in the immediate vicinity of the nerve membrane (Narahashi and Yamasaki, 1960a). Electron microscope observations support this view (Smith and Treherne, 1963).

The negative after-potential is liable to be affected by various experimental conditions. It is augmented and prolonged by barium ions (Narahashi, 1961), DDT (Narahashi and Yamasaki, 1960b, c), allethrin (Narahashi, 1962a, b), high calcium (Narahashi and Yamasaki, 1960a), or high temperature (Boistel, 1960, 1962a, b; Narahashi, 1963). In such cases, large negative after-potentials are not necessarily attributable to the potassium accumulation. For example, the suppression of the potassium-activation mechanism is involved in the case of DDT (Narahashi and Yamasaki, 1960b, c), and the accumulation of an unknown depolarizing substance is suggested in the case of allethrin (Narahashi, 1962a, b).

II. Pharmacological Aspects of Nervous Functions

A. *Classification of the Conditions that Affect the Nervous Functions*

The properties of the nerve membrane can be altered by a variety of means. The following is a classification of the conditions based on the mechanisms whereby the membrane properties are changed:

(a) Those substances which change the resting and/or action potential by simply altering the e.m.f. of the membrane. High potassium and low sodium are typical examples.

(b) Those substances which combine with or are absorbed into the nerve membrane, thereby altering the membrane properties. This group may be further divided into two: (1) those which hinder membrane potential changes produced by various means and block conduction without affecting the resting potential, i.e. stabilizers (Shanes, 1958a, b) such as high calcium and cocaine; (2) those which enhance or accelerate membrane potential changes, and may cause repetitive activity, i.e. labilizers (Shanes, 1958a, b) such as DDT and veratrine.

(c) Those which block excitability through the inhibition of energy metabolism, e.g. rotenone (Fukami, Nakatsugawa and Narahashi, 1959).

The group (a) above has already been described. The group (b) is of special interest, because ions and drugs belonging to this group may be used as tools to explore the mechanisms of excitation. Here, some examples will be presented in which attempts have been made to account for the action of drugs in terms of membrane conductance changes.

B. *Possible Approaches*

It has become apparent that at least three factors govern excitability: (1) the ability of the membrane to undergo a sodium conductance increase upon depolarization, the "sodium-activation mechanism"; (2) the ability of the membrane to undergo a sodium conductance decrease during sustained depolarization, the "sodium-inactivation mechanism"; (3) the ability of the membrane to undergo a potassium conductance increase upon depolarization, the "potassium-activation mechanism". There is also evidence that, at least in certain nerves and muscles, the "potassium-inactivation mechanism" is operative (Frankenhaeuser and Waltman, 1959; Grundfest, 1960; Nakajima, Iwasaki and Obata, 1962).

In order to elucidate the mechanism of action of the agents which affect the membrane, it is necessary to interpret the action on the nerve membrane in terms of the above three, or possibly four, factors. The ideal way of doing this would no doubt be to use the voltage-clamp technique, but so far no attempts have been made to apply this to insect nerves. This technique requires space clamping, which was formerly only possible by introducing longitudinal electrodes into an axon; and is very difficult, if not impossible, in the small insect axons. However, it has now become possible to establish a space-clamp condition by means of the "sucrose-gap" method without having any electrode inside the axon (Julian, Moore and Goldman, 1962a). This method was successfully used in voltage-clamp experiments with lobster giant axons which were only twice as large in diameter as cockroach giant axons (Fig. 7) (Julian *et al.*, 1926b; Narahashi *et al.*, 1964). It is hoped that this method will be applicable to insect giant axons in the near future.

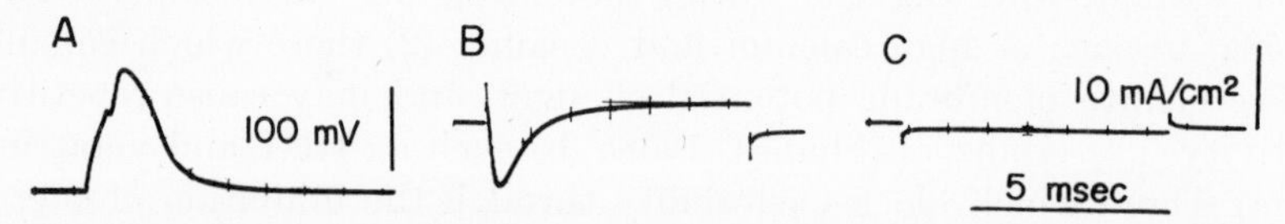

FIG. 7. Voltage-clamp experiments in a lobster giant axon by means of the sucrose-gap method. A, voltage is not clamped. An action potential is produced by an outward current. The critical depolarization for firing is high because of hyperpolarization by sucrose. B, current recording when the membrane is depolarized by 90 mV from the holding membrane potential of −115 mV. Upward deflexion denotes outward current. Note the initial and transient inward current and the delayed and sustained outward current. C, current recording when the membrane is hyperpolarized by 40 mV. Note the simple inward current. Preparation 63–116Af. Temperature 10° C.

It is also possible, however, to analyse, by means of the conventional intracellular microelectrodes, changes in the above four factors by the action of membrane-attacking agents. Since the maximum rate of rise of the action potential is proportional to the inward sodium current at that moment in the propagating action potential (Narahashi, 1961), this parameter can be used as an indication of the sodium-activation mechanism. Delayed rectification revealed by cathodal current is known to be an expression of the sustained potassium conductance rise during depolarization (Hodgkin *et al.*, 1949; Yamasaki and Narahashi, 1959b), and can be used as a measure of the potassium-activation mechanism. The potassium-activation is also manifested as prolonged depolarizing responses in sodium-free potassium-rich media when the membrane potential is brought to appropriately high levels (Fig. 8) (Ooyama and Wright, 1962; Wright and Tomita, 1962). We have for the moment no direct way of estimating the sodium-inactivation mechanism, but the activity of the mechanism may be inferred to some extent from observation of the maximum rate of fall of the action potential and of delayed rectification, because the maximum rate of fall is determined both by the sodium inactivation and by the potassium

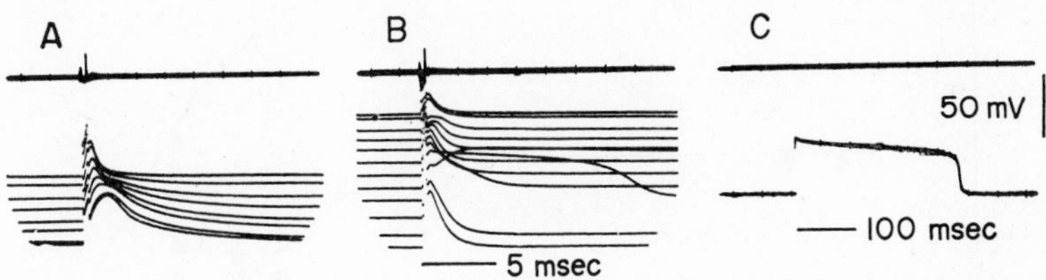

FIG. 8. Responses of a cockroach giant axon in Na-free 3·1 mM–K choline Ringer (A) and in Na-free 40 mM–K Ringer (B and C). A brief cathodal current is applied to the membrane which has been hyperpolarized to various levels by sustained anodal currents. Upper tracings show the zero potential level. Superimposed (A and B) and single (C) records. Note the appearance of prolonged depolarizing responses in high-K (B and C) and its absence in low-K (A). The responses are regarded as indicating the potassium-activation mechanism. Preparation 63–816Aa. Temperature 34° C.

activation. The potassium-inactivation mechanism, when present, may be estimated by observing the time course of change in delayed rectification or the current-voltage relations (Grundfest, 1960).

C. *Allethrin and Tetrodotoxin*

Allethrin, a derivative of pyrethrins, has at least three distinct effects on the cockroach giant axons: (1) increase in negative after-potential;

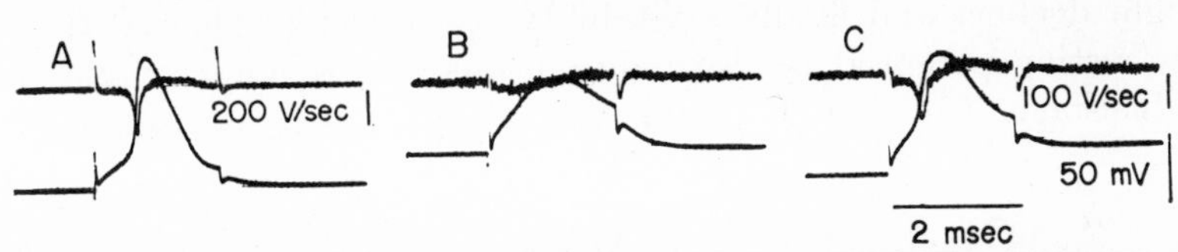

FIG. 9. The effect of allethrin 10^{-6} g/ml on the action potential of the cockroach giant axon. Upper tracings represent the rate of rise (downward deflexion) and the rate of fall (upward deflexion) of the action potential, and the zero potential level. Lower tracings represent the action potential produced by a pulse of cathodal current. A, before allethrin; B, 12 minutes after allethrin, note the slight depolarization and the disappearance of the action potential; C, as in B, but the action potential is restored by anodal polarization. Preparation 64–130Aa. Temperature 18° C.

(2) repetitive after-discharge; (3) conduction block. The first two effects have already been analysed and reported elsewhere (Narahashi, 1962a, b). Here the third effect will be described. Tetrodotoxin is an active principle of puffer poison, and in lobster axons and in frog muscle

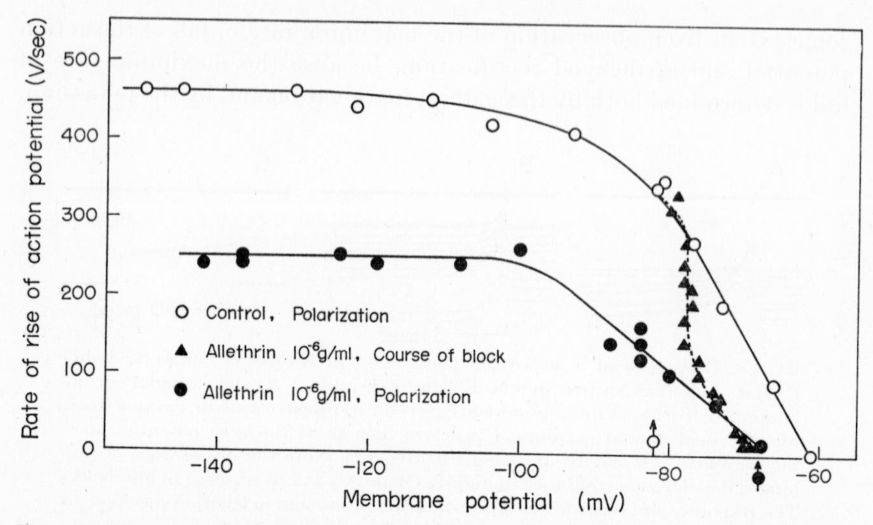

FIG. 10. The maximum rate of rise of the action potential of a cockroach giant axon plotted against the membrane potentials displaced by polarizing currents before (○) and 18 minutes after (●) allethrin 10^{-6} g/ml (sodium-inactivation curves). The arrows indicate the resting potentials. The values of the rate of rise during the course of blockade after allethrin are also plotted against the respective membrane potentials (▲). Preparation 64–128A. Temperature 23° C.

fibres has been known to block conduction through the selective inhibition of the sodium-activation mechanism (Narahashi, *et al.*, 1960, 1964). This is used here for comparison.

After exposure to allethrin at a concentration of 10^{-6} g/ml, the spike height declines and finally excitability is completely blocked (Fig. 9). The resting potential is also reduced, but this is not the sole cause of blockage. The decline of the spike height is greater than would

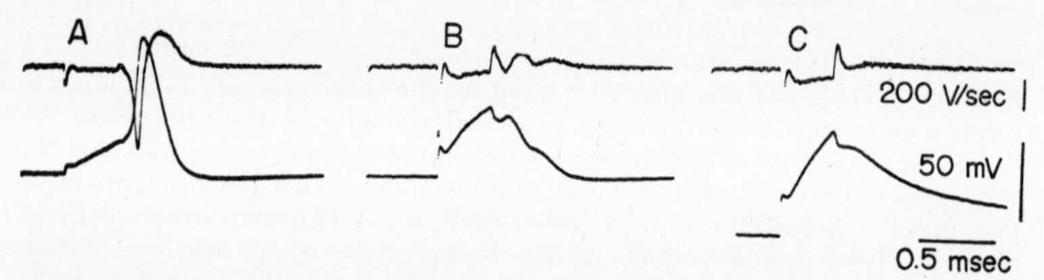

FIG. 11. The effect of tetrodotoxin 3×10^{-8} g/ml on the action potential of the cockroach giant axon. Upper tracings represent the rate of rise (downward deflexion) and the rate of fall (upward deflexion) of the action potential, and the zero potential level. Lower tracings represent the action potential produced by a pulse of cathodal current. A, before tetrodotoxin; B, 5 minutes after tetrodotoxin, note the drastic reduction of the action potential; C, as in B, showing the inability of anodal polarization to restore the action potential. Preparation 63–822Ba. Temperature 31·5° C.

be expected when the membrane is depolarized by cathodal current (Fig. 10). The block after tetrodotoxin (10^{-7}–10^{-8} g/ml) is not accompanied by any significant change in resting potential (Figs. 11 and 12).

The maximum rates of rise and fall of the action potential decreases at about the same rate during blockage after treating with allethrin. This situation is in sharp contrast with the blockage obtained after treatment with Na-free media or tetrodotoxin, in which the rate of rise

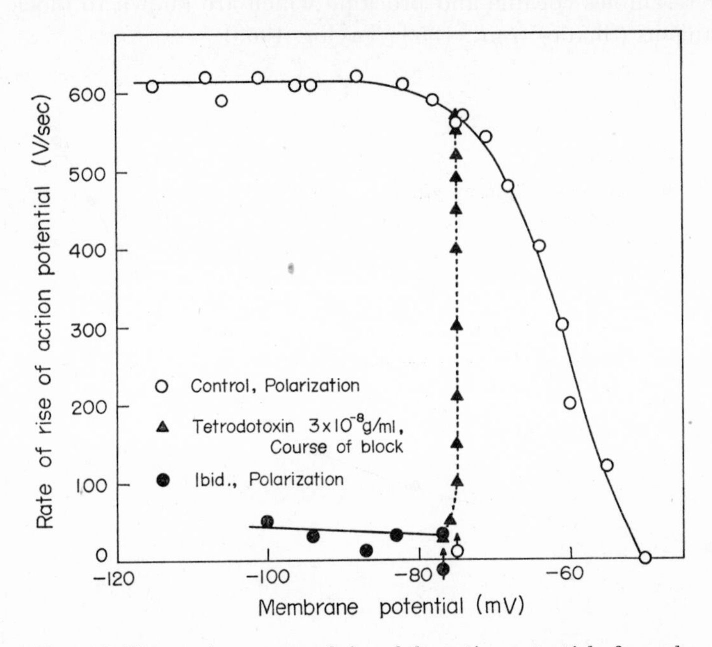

FIG. 12. The maximum rate of rise of the action potential of a cockroach giant axon plotted against the membrane potentials displaced by polarizing currents before (○) and 5 minutes after (●) tetrodotoxin 3×10^{-8} g/ml (sodium-inactivation curves). The arrows indicate the resting potentials. The values of the rate of rise during the course of blockade after tetrodotoxin are also plotted against the respective membrane potentials (▲). Preparation 63–822Ba. Temperature 31·5° C.

decreases much faster than does the rate of fall. This is to be expected since the sodium-activation mechanism cannot manifest its activity in Na-free media because of the lack of sodium, and is completely and very selectively blocked by tetrodotoxin. These considerations exclude the possibility that allethrin blocks the sodium-activation mechanism selectively. Delayed rectification is suppressed by treatment with allethrin (Fig. 13). This strongly suggests that the potassium-activation mechanism is inhibited to some extent. So far we do not have sufficient data to infer the situation with respect to the sodium-inactivation mechanism.

Thus it may be concluded that allethrin, at least at its blocking concentrations, inhibits both sodium- and potassium-activation mechanisms. The situation is in contrast with that of tetrodotoxin which inhibits the sodium-activation mechanism selectively without affecting the potassium-activation mechanism (Narahashi *et al.*, 1960, 1964), and rather resembles cocaine and procaine which are known to block both mechanisms (Shanes *et al.*, 1959; Taylor, 1959).

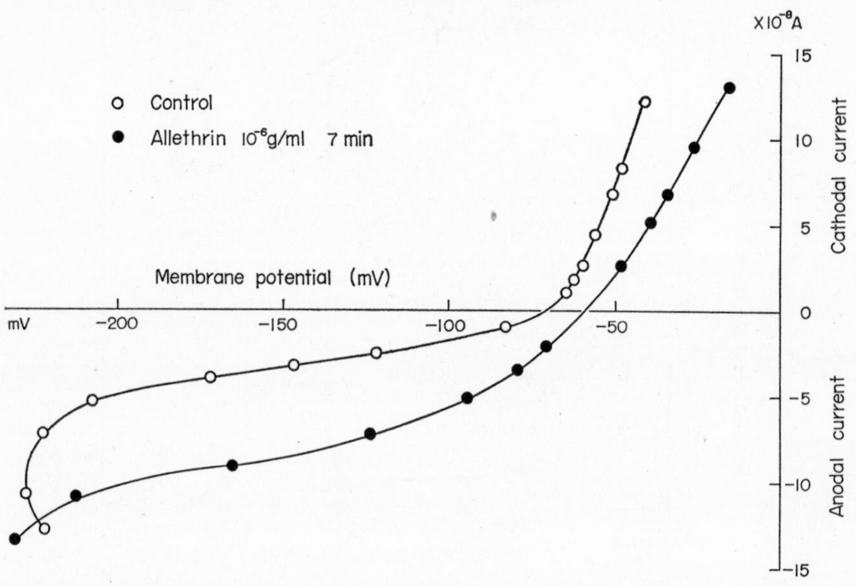

FIG. 13. Current-voltage relations of a cockroach giant axon before (○) and 7 minutes after (●) allethrin 10^{-6} g/ml. Note that delayed rectification as indicated by an upward curvature is less marked after allethrin. Preparation 64–210Ba. Temperature 21° C.

Another important difference between allethrin and tetrodotoxin is that anodal polarization can relieve the allethrin block (Figs. 9 and 10) but cannot relieve the tetrodotoxin block (Figs. 11 and 12). This suggests different sites of action of these two agents in the membrane (cf. Narahashi, 1964).

III. Molecular Aspects of Nervous Functions

It has been well established that the thin nerve membrane is the major site where various excitation phenomena take place (e.g. Villegas *et al.*, 1963). Calcium is known to bind with membrane molecules such as phospholipid, cholesterol and protein (cf. Brink, 1954; Tobias and

Nelson, 1959). There is also the widely accepted view that upon depolarization, calcium is displaced from its binding site, thereby increasing the membrane conductance to sodium and potassium (e.g. Frankenhaeuser and Hodgkin, 1957; Koketsu *et al.*, 1963; Narahashi, 1964; Shanes, 1958a, b, 1960; Tobias, 1958).

Calcium also plays an important role in excitability in the cockroach giant axons. Increasing external calcium concentration raised the resting potential by a few millivolts. The action potential is increased in height but decreased in its maximum rate of rise. However, one of the most important changes produced by high calcium is that the membrane

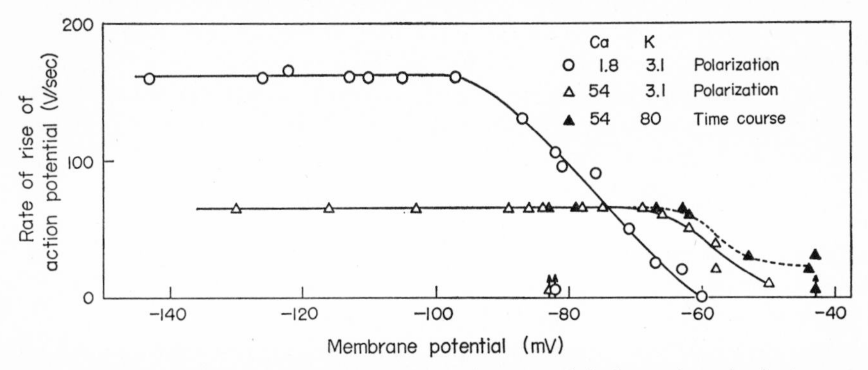

FIG. 14. The maximum rate of rise of the action potential of a cockroach giant axon plotted against the membrane potentials displaced by polarizing currents in 1·8 mM–Ca Ringer (○) and in 54 mM–Ca Ringer (△) (sodium-inactivation curves). The arrows indicate the resting potentials. The values of the rate of rise after exposure to 54 mM–Ca 80 mM–K Ringer are also plotted against the respective membrane potentials (▲). The sodium concentration is kept constant throughout. Note the shift of the curve toward lower membrane potential in the high-Ca regardless of the potassium concentration. Preparation 64-219Aa. Temperature 20° C.

could be greatly depolarized by cathodal current or by high potassium before losing excitability (Fig. 14). In other words, the curve relating the maximum rate of rise of the action potential or the inward sodium current to the membrane potential is shifted along the potential axis to a lower membrane potential; this curve is called the "sodium-inactivation curve", because it represents the degree of sodium inactivation as a function of membrane potential (Hodgkin and Huxley, 1952). Addition of barium also causes the curve to be shifted to a lower membrane potential.

Lowering calcium concentration to one-tenth normal or to zero causes progressive depolarization and block. It is, however, possible to restore excitability by anodal hyperpolarization. In this case, the sodium-inactivation curve is shifted to a higher membrane potential.

Similar shift of the sodium-inactivation curve has been observed with squid or lobster giant axons (Frankenhaeuser and Hodgkin, 1957; Narahashi, 1964), or with the Purkinje fibres of mammals (Weidmann, 1955).

Two important conclusions are derived from the above results. First of all, as in many other excitable tissues, the presence of calcium in the external medium of cockroach giant axons is essential for excitability to persist. It is reasonable to assume that the calcium in the membrane is in a state of equilibrium with the external calcium. This and much other available evidence (e.g. Koketsu and Miyamoto, 1961a, b, Narahashi, 1964; Shanes and Bianchi, 1960) leads us to conclude that the integrity of the membrane molecular structure in which calcium works as a bridge or binder must be maintained for excitability to persist. Secondly, the results clearly account for the so-called K–Ca antagonism, a phenomenon which has long been known. The inability of high potassium to block excitability in the presence of high calcium is clearly shown by a shift in the sodium-inactivation curve.

We still fall far short of the ultimate goal of explaining various neurophysiological problems in physico-chemical or molecular terms. We are also far from being able to draw a map or scheme in which the network of impulse pathways can adequately describe any behaviour or functions of insects. It is to be hoped that learning or conditioning may be accounted for in terms of a synaptic sequence of events such as facilitation and inhibition, which are in turn manifestations of the changes at molecular level in the presynaptic nerve terminals and at the postsynaptic membrane. In the applied field of entomology, clarification of insecticidal actions in terms of molecular interactions between insecticides and the nerve membrane structure may also help in the search for new insecticides.

IV. Summary

It has now become possible to interpret membrane potential and excitation in cockroach giant axons by the ionic theory. The agents which exert their effects on excitability through interactions with the nerve membrane are of special interest because of their possible use as tools to explore the excitation mechanism. Electrophysiological analyses on the blocking actions of allethrin and of tetrodotoxin are described. Attempts are also made to clarify the role of calcium in excitability. Available evidence strongly supports the notion that the integrity of molecules, especially of calcium, in the membrane must be maintained for excitability to persist.

REFERENCES

Bernstein, J. (1912). "Elektrophysiologie". 215 pp. Fr. Vieweg u. Sohn, Braunschweig.

Boistel, J. (1959). Quelques caractéristiques électriques de la membrane de la fibre nerveuse au repos d'un insecte (*Periplaneta americana*). *C. R. Soc. Biol., Paris* **153,** 1009–1013.

Boistel, J. (1960). "Caractéristiques fonctionnelles des fibres nerveuses et des récepteurs tactiles et olfactifs des insectes". 147 pp. Librairie Arnette, Paris.

Boistel, J. (1962a). Analyse des effets produits par une elevation de temperature sur le potentiel consecutif de la fibre géante de blatte (*Periplaneta americana*). *Excerpta med.,* Int. Congr. Ser. No. 48, 778.

Boistel, J. (1962b). Interpretation des modifications du potentiel consecutif de la fibre géante de blatte sous l'effet de la temperature. *Symp. Genet. Biol. Italica* **10,** 278–290.

Boistel, J., and Coraboeuf, É. (1954). Potentiel de membrane et potentiels d'action de nerf d'insecte recueillis à l'aide de microélectrodes intracellulaires. *C. R. Acad. Sci. Paris* **238,** 2116–2118.

Brink, F. (1954). The role of calcium ions in neural processes. *Pharmacol. Rev.* **6,** 243–298.

Cole, K. S. (1949). Dynamic electrical characteristics of the squid axon membrane. *Arch. Sci. Physiol.* **3,** 253–258.

Cole, K. S., and Hodgkin, A. L. (1939). Membrane and protoplasm resistance in the squid giant axon. *J. gen. Physiol.* **22,** 671–687.

Cole, K. S., and Moore, J. W. (1960). Ionic current measurements in the squid giant axon membrane. *J. gen. Physiol.* **44,** 123–167.

Curtis, H. J., and Cole, K. S. (1940). Membrane action potentials from the squid giant axon. *J. cell. comp. Physiol.* **15,** 147–157.

Curtis, H. J., and Cole, K. S. (1942). Membrane resting and action potentials from the squid giant axon. *J. cell. comp. Physiol.* **19,** 135–144.

Frankenhaeuser, B., and Hodgkin, A. L. (1957). The action of calcium on the electrical properties of squid axons. *J. Physiol.* **137,** 218–244.

Frankenhaeuser, B., and Waltman, B. (1959). Membrane resistance and conduction velocity of large myelinated nerve fibre from *Xenopus laevis*. *J. Physiol.* **148,** 677–682.

Fukami, J., Nakatsugawa, T., and Narahashi, T. (1959). The relation between chemical structure and toxicity in rotenone derivatives. *Japan. J. appl. Ent. Zool.* **3,** 259–265.

Grundfest, H. (1960). Ionic mechanisms in electrogenesis. *Ann. N. Y. Acad. Sci.* **94,** 405–457.

Hagiwara, S., and Saito, N. (1959a). Voltage-current relations in nerve cell membrane of *Onchidium verruculatum*. *J. Physiol.* **148,** 161–179.

Hagiwara, S., and Saito, N. (1959b). Membrane potential change and membrane current in supramedullary nerve cell of puffer. *J. Neurophysiol.* **22,** 204–221.

Hodgkin, A. L. (1947). The membrane resistance of a non-medullated nerve fibre. *J. Physiol.* **106,** 305–318.

Hodgkin, A. L. (1951). The ionic basis of electrical activity in nerve and muscle. *Biol. Rev.* **26,** 339–409.

Hodgkin, A. L. (1958). Ionic movements and electrical activity in giant nerve fibres. *Proc. roy. Soc.* B. **148,** 1–37.

Hodgkin, A. L., and Huxley, A. F. (1939). Action potentials recorded from inside a nerve fibre. *Nature, Lond.* **144,** 710–711.

Hodgkin, A. L., and Huxley, A. F. (1945). Resting and action potentials in single nerve fibres. *J. Physiol.* **104,** 176–195.

Hodgkin, A. L., and Huxley, A. F. (1952). The dual effect of membrane potential on sodium conductance in the giant axon of *Loligo. J. Physiol.* **116,** 497–506.

Hodgkin, A. L., and Katz, B. (1949). The effect of sodium ions on the electrical activity of the giant axon of the squid. *J. Physiol.* **108,** 37–77.

Hodgkin, A. L., and Rushton, W. A. H. (1946). The electrical constants of a crustacean nerve fibre. *Proc. roy. Soc.* B. **133,** 444–479.

Hodgkin, A. L., Huxley, A. F., and Katz, B. (1949). Ionic currents underlying activity in the giant axon of the squid. *Arch. Sci. Physiol.* **3,** 129–150.

Hodgkin, A. L., Huxley, A. F., and Katz, B. (1952). Measurement of current-voltage relations in the membrane of the giant axon of *Loligo. J. Physiol.* **116,** 424–448.

Huxley, A. F., and Stämpfli, R. (1951). Effect of potassium and sodium on resting and action potentials of single myelinated nerve fibres. *J. Physiol.* **112,** 496–508.

Julian, F. J., Moore, J. W., and Goldman, D. E. (1962a). Membrane potentials of the lobster giant axon obtained by use of the sucrose-gap technique. *J. gen. Physiol.* **45,** 1195–1216.

Julian, F. J., Moore, J. W., and Goldman, D. E. (1962b). Current-voltage relations in the lobster giant axon membrane under voltage clamp conditions. *J. gen. Physiol.* **45,** 1217–1238.

Koketsu, K., and Miyamoto, S. (1961a). Release of calcium-45 from frog nerves during electrical activity. *Nature, Lond.* **189,** 402–403.

Koketsu, K., and Miyamoto, S. (1961b). Significance of membrane calcium in calcium-free and potassium-rich media. *Nature, Lond.* **189,** 403–404.

Koketsu, K., Nishi, S., and Soeda, H. (1963). Effects of calcium ions on prolonged action potentials and hyperpolarizing responses. *Nature, Lond.* **200,** 786–787.

Ling, G., and Gerard, R. W. (1950). External potassium and the membrane potential of single muscle fibres. *Nature, Lond.* **165,** 113–114.

Marmont, G. (1949). Studies on the axon membrane. *J. cell. comp. Physiol.* **34,** 351–382.

Nakajima, S., Iwasaki, S., and Obata, K. (1962). Delayed rectification and anomalous rectification in frog's skeletal muscle membrane. *J. gen. Physiol.* **46,** 97–115.

Narahashi, T. (1961). Effect of barium ions on membrane potentials of cockroach giant axons. *J. Physiol.* **156,** 389–414.

Narahashi, T. (1962a). Effect of the insecticide allethrin on membrane potentials of cockroach giant axons. *J. cell. comp. Physiol.* **59,** 61–65.

Narahashi, T. (1962b). Nature of the negative after-potential increased by the insecticide allethrin in cockroach giant axons. *J. cell. comp. Physiol.* **59,** 67–76.

Narahashi, T. (1963). The properties of insect axons. *In* "Advances in Insect Physiology" (J. W. L. Beament, J. E. Treherne and V. B. Wigglesworth, eds.), Vol. 1, pp. 175–256. Academic Press, London and New York.

Narahashi, T. (1964). Restoration of action potential by anodal polarization in lobster giant axons. *J. cell. comp. Physiol.* **64,** 73–96.

Narahashi, T., and Yamasaki, T. (1960a). Mechanism of the after-potential production in the giant axons of the cockroach. *J. Physiol.* **151**, 75–88.

Narahashi, T., and Yamasaki, T. (1960b). Mechanism of increase in negative after-potential by dicophanum (DDT) in the giant axons of the cockroach. *J. Physiol.* **152**, 122–140.

Narahashi, T., and Yamasaki, T. (1960c). Behaviors of membrane potential in the cockroach giant axons poisoned by DDT. *J. cell. comp. Physiol.* **55**, 131–142.

Narahashi, T., Moore, J. W., and Scott, W. R. (1964). Tetrodotoxin blockage of sodium conductance increase in lobster giant axons. *J. gen. Physiol.* **47**, 965–974.

Narahashi, T., Deguchi, T., Urakawa, N., and Ohkubo, Y. (1960). Stabilization and rectification of muscle fiber membrane by tetrodotoxin. *Amer. J. Physiol.* **198**, 934–398.

Nastuk, W. L., and Hodgkin, A. L. (1950). The electrical activity of single muscle fibers. *J. cell. comp. Physiol.* **35**, 39–74.

Ooyama, H., and Wright, E. B. (1962). Activity of potassium mechanism in single Ranvier node during excitation. *J. Neurophysiol.* **25**, 67–93.

Shanes, A. M. (1958a). Electrochemical aspects of physiological and pharmacological action in excitable cells. Part I. The resting cell and its alteration by extrinsic factors. *Pharmacol. Rev.* **10**, 59–164.

Shanes, A. M. (1958b). Electrochemical aspects of physiological and pharmacological action in excitable cells. Part II. The action potential and excitation. *Pharmacol. Rev.* **10**, 165–273.

Shanes, A. M. (1960). Mechanism of change in permeability in living membranes. *Nature, Lond.* **188**, 1209–1210.

Shanes, A. M., and Bianchi, C. P. (1960). Radiocalcium release by stimulated and potassium-treated sartorius muscle of the frog. *J. gen. Physiol.* **43**, 481–493.

Shanes, A. M., Freygang, W. H., Grundfest, H., and Amatniek, E. (1959). Anesthetic and calcium action in the voltage clamped squid giant axon. *J. gen. Physiol.* **42**, 793–802.

Smith, D. S., and Treherne, J. E. (1963). Functional aspects of the organization of the insect nervous system. *In* "Advances in Insect Physiology" (J. W. L. Beament, J. E. Treherne and V. B. Wigglesworth, eds.), Vol. I, pp. 401–484. Academic Press, London and New York.

Tasaki, I., and Hagiwara, S. (1957). Demonstration of two stable potential states in the squid giant axon under tetraethylammonium chloride. *J. gen. Physiol.* **40**, 859–885.

Taylor, R. E. (1959). Effect of procaine on electrical properties of squid axon membrane. *Amer. J. Physiol.* **196**, 1071–1078.

Tobias, J. M. (1948). Potassium, sodium and water interchange in irritable tissues and haemolymph of an omnivorous insect, *Periplaneta americana*. *J. cell. comp. Physiol.* **31**, 125–142.

Tobias, J. M. (1958). Experimentally altered structure related to function in the lobster axon with an extrapolation to molecular mechanisms in excitation. *J. cell. comp. Physiol.* **52**, 89–125.

Tobias, J. M., and Bryant, S. H. (1955). An isolated giant axon preparation from the lobster nerve cord. Dissection, physical structure, trans-surface potentials and microinjection. *J. cell. comp. Physiol.* **46**, 163–182.

Tobias, J. M., and Nelson, P. G. (1959). Structure and function in nerve. *In* "A Symposium on Molecular Biology" (R. E. Zirkle, ed.), pp. 248–265. University of Chicago Press, Chicago.

Treherne, J. E. (1961). Sodium and potassium fluxes in the abdominal nerve cord of the cockroach, *Periplaneta americana* L. *J. exp. Biol.* **38,** 315–322.

Villegas, R., Villegas, L., Giménez, M., and Villegas, G. M. (1963). Schwann cell and axon electrical potential differences. Squid nerve structure and excitable membrane location. *J. gen. Physiol.* **46,** 1047–1064.

Weidmann, S. (1955). Effects of calcium ions and local anaesthetics on electrical properties of Purkinje fibres. *J. Physiol.* **129,** 568–582.

Wright, E. B., and Tomita, T. (1962). Separation of sodium and potassium ion carrier systems in crustacean motor axon. *Amer. J. Physiol.* **202,** 856–864.

Yamasaki, T., and Narashashi, T. (1959a). The effects of potassium and sodium ions on the resting and action potentials of the cockroach giant axon. *J. Insect Physiol.* **3,** 146–158.

Yamasaki, T., and Narahashi, T. (1959b). Electrical properties of the cockroach giant axon. *J. Insect Physiol.* **3,** 230–242.

The Chemical Environment of the Insect Central Nervous System

J. E. TREHERNE

A.R.C. Unit of Insect Physiology, Department of Zoology, University of Cambridge

The function of nerve cells ultimately depends upon the chemical composition of the fluid immediately surrounding them. The nature of this extracellular environment in the insect central nervous system is of especial interest in view of the very specialized composition of the haemolymph of species from orders such as the Lepidoptera, Phasmida, Hymenoptera and Coleoptera. In these largely phytophagous insects the sodium content of the haemolymph is relatively low, often consisting of only a few mmoles/l, and is frequently much lower than the level of potassium or magnesium ions, which are of a higher order of concentration than in vertebrate body fluids (cf. Wyatt, 1961; Shaw and Stobbart, 1963; Sutcliffe, 1963). Such a peculiar environment presents certain difficulties in the interpretation of nerve function according to the classical theory of the transmission of the nerve impulse (cf. Hodgkin, 1951, 1958), which, as the preceding paper has emphasized, seems to be applicable to the giant axons of at least one insect species (see also Narahashi, 1963). It is the purpose of this contribution to summarize our present state of knowledge of the exchange and distribution of ions and molecules in the tissues of the central nervous system and to integrate it with what is known of the electrical properties of insect nerve cells.

Despite the differences in ionic composition of the haemolymph of an insect such as *Periplaneta americana* and a phytophagous species such as *Carausius morosus*, the concentrations of inorganic ions in the central nervous system appear to be similar in many respects. Thus, although the sodium level of the haemolymph of the latter species is an order lower than that of *Periplaneta*, the gross tissue concentrations of this cation are approximately similar to these two insects (Treherne, 1961a, 1964b).

Recent experiments using radioisotopes have demonstrated rapid exchanges of inorganic ions and molecules between the haemolymph and the central tissues in both *Periplaneta* and *Carausius* (Treherne, 1961a, b, c; 1964a, b). The uptake of sodium was found to occur at approximately the same rate in the nerve cords of these insects, despite the fact that this ion is of an order more concentrated in the haemolymph of *Periplaneta* than in the phytophagous species (Fig. 1).

The exchanges of inorganic ions occurring within the central nervous tissues is most clearly seen in experiments in which the nerve cords are

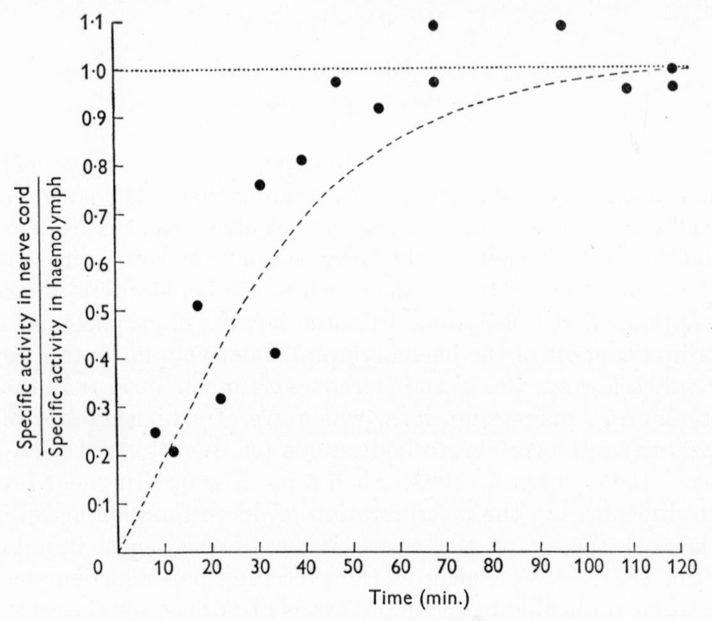

FIG. 1. Changes in the specific activity of ^{22}Na in the nerve cord of *Carausius morosus*, relative to that in the haemolymph, following the injection of the radioisotope into the haemolymph. The broken line shows the equivalent changes in specific activity measured in the abdominal nerve cord of *Periplaneta americana*. (Data from Treherne, 1961a, 1964b.)

made radioactive by the injection of the appropriate isotope into the haemolymph, the subsequent efflux being studied by washing the isolated preparation in non-radioactive physiological solution (Fig. 2). In the nerve cords of both *Periplaneta* and *Carausius* the escape of inorganic ions was found to occur as a two-stage process, an initial rapid phase eventually giving way to a slow exponential efflux (Treherne, 1961d, 1964b). The evidence which has accumulated indicates that the initial rapidly exchanging portion of the efflux curve is, in fact, the extracellular ion fraction of the central nervous tissues. This identification is based on the following observations:

1. The greater part of the rapidly exchanging ion fraction was not associated with the nerve sheath, for in *Periplaneta* this component persisted in desheathed preparations (Treherne, 1962), while in *Carausius* the radioactivity remaining as a result of a brief exposure to radiosodium formed only a small proportion of the total rapidly exchanging fraction (Treherne, 1964b).

2. In both species the escape of the rapidly exchanging fraction was unaffected by the presence of metabolic inhibitors, although the efflux

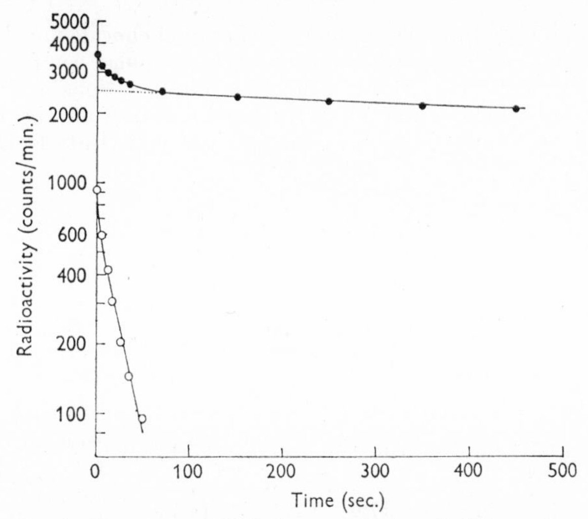

FIG. 2. The escape of ^{22}Na from a nerve cord of *Carausius morosus* when washed in non-radioactive physiological solution (closed circles). The nerve cord was made radioactive by the injection of the radioisotope into the haemolymph 4 hr. before the commencement of the experiment. The fast component of the main curve (open circles) was obtained by subtraction from the straight line extrapolated to zero time. (Treherne, 1964b.)

of the slowly exchanging sodium was reduced under these circumstances (Treherne, 1961d, 1964b).

3. The efflux of tritiated water was found to occur as a two-stage process as in the case of the inorganic ions (Treherne, 1962).

4. The rapidly exchanging tritiated water fraction was of a similar magnitude to the measured inulin space in the nerve cord of *Periplaneta* (Treherne, 1962).

5. Electronmicrographs of *Periplaneta* nerve cord demonstrated the existence of an appreciable extracellular system sufficient to accommodate the inulin molecules used in the previous experiments (Smith and Treherne, 1963).

If the values for the inulin space or the rapidly exchanging tritiated water fraction are used in conjunction with those for the rapidly exchanging ion fractions, it is possible to arrive at an estimate for the ionic composition of the extracellular fluid. In both *Periplaneta* (Treherne, 1962) and *Carausius* (Treherne, 1964b) the estimated concentrations of the extracellular ions were found to be markedly different from those in the haemolymph (Tables I and II).

TABLE I

The distribution of sodium, potassium, calcium and choride ions between the external medium and the extracellular and intracellular components of the nerve cord of *Periplaneta*. These estimated concentrations are based on a volume of extracellular water of 21·6%, calculated from the rapidly exchanging tritiated water fraction in this preparation. (Data from Treherne, 1961a, 1962.)

Ion	External concentration mmoles/l	Estimated extracellular concentration mmoles/l	Estimated intracellular concentration mmoles/l	Ion ratio
Na	157·0	283·6	67·2	$\frac{Na_{out}}{Na_{extracellular}} = 0{\cdot}55$
K	12·3	17·1	225·1	$\frac{K_{out}}{K_{extracellular}} = 0{\cdot}71$
Ca	4·5	17·6	14·7	$\frac{(Ca_{out})^{\frac{1}{2}}}{(Ca_{extracellular})^{\frac{1}{2}}} = 0{\cdot}51$
Cl	184·0	106·7	—	$\frac{Cl_{extracellular}}{Cl_{out}} = 0{\cdot}58$

In the nerve cord of *Periplaneta* the high extracellular concentrations of the cations and the relatively low level of chloride ions is quite consistent with the hypothesis that the distribution is governed by a Donnan equilibrium with the haemolymph (Treherne, 1962). The molecules containing the free anion groups in the extracellular system have not been conclusively identified, although it is possible that the acid mucopolysaccharide demonstrated in the extracellular spaces of cockroach ganglia (Ashhurst, 1961) may contribute to the demonstrated Donnan equilibrium.

It is clear from these considerations that the removal of the nerve sheath results in extensive changes in the composition of the extracellular fluid, due to the disruption of the Donnan equilibrium with the haemolymph. This effect seems to result not only from a dispersal of molecules containing free anion groups, but also from a large increase in the volume of the extracellular fluid in desheathed preparations (Treherne, 1962). It seems likely that these alterations in the extracellular environment of the nerve cells could contribute to the enhanced rates of depolarization observed at elevated potassium concentrations in desheathed preparations (Hoyle, 1953; Twarog and Roeder, 1956), rather than to any properties of the nerve sheath of the central nervous system as a significant diffusion barrier.

TABLE II

The calculated concentrations of the rapidly exchanging ion fractions in the nerve cord of *Carausius*. The estimates of the extracellular concentration are based on the measured inulin space of the nerve cord (Treherne, 1964b).

Ion	Type of experiment	Concentration in haemolymph or external solution mmoles/l	Calculated extracellular concentration mmoles/l	Ion ratio
Na	*in vivo*	20·1	212·4†	$\frac{Na_{extracellular}}{Na_{out}}$ = 10·6
	in vitro	15·0	78·7‡	= 5·2
	in vitro (poisoned)	15·0	18·5‡	= 1·2
K	*in vitro*	18·0	67·1‡	$\frac{K_{extracellular}}{K_{out}}$ = 3·7
	in vitro (poisoned)	18·0	72·5‡	= 4·0
Ca	*in vitro*	7·5	14·4‡	$\left(\frac{Ca_{extracellular}}{Ca_{out}}\right)^{\frac{1}{2}}$ = 1·4
	in vitro (poisoned)	7·5	15·0‡	
Mg	*in vitro*	40·0	94·4‡	$\left(\frac{Mg_{extracellular}}{Mg_{out}}\right)^{\frac{1}{2}}$ = 1·4
Cl	*in vitro*	133·0	80·7–214·9‡	

† Based on *in vivo* inulin space of 91·8 ml/Kg tissue.
‡ Based on *in vitro* inulin space of 146·1 ml/Kg tissue.

As in *Periplaneta* the concentrations of the major cations in the rapidly exchanging fraction in the nerve cord of *Carausius* were found to be very different from their levels in the haemolymph (Treherne, 1964b). Although the concentrations of the major cations in the extracellular fluid exceed those in the haemolymph, their distribution does not, as in the cockroach, conform to a simple Donnan equilibrium (Table II). In particular, the very high concentration ratios for the monovalent cations, as compared with the divalent ones, are very difficult to explain simply on the basis of unspecific fixed or indiffusible free anion groups in the extracellular system. It would be expected, for example, that in the event of a simple Donnan effect the divalent cation levels would exceed those for the monovalent cations.

The steep concentration gradient for sodium ions between the extracellular fluid and the haemolymph in the nerve cord of *Carausius* was found to be drastically reduced when the preparation was loaded with the radioisotope *in vitro*, and to be virtually abolished in the presence of dilute cyanide or dinitrophenol (Treherne, 1964b). The high extracellular concentration of sodium in the central nervous tissues of this species appears to result from a secretion of this ion from the haemo-

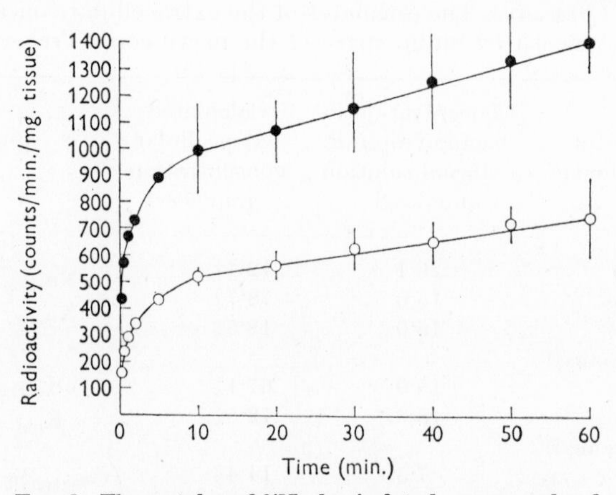

FIG. 3. The uptake of ^{24}Na by isolated nerve cords of *Carausius morosus* soaked in radioactive physiological solution. The closed circles represent the uptake in normal solution; the open circles, that measured in a solution containing 0·5 mmole/l 2 : 4 dinitrophenol (Treherne, 1964b).

lymph into the extracellular system, for in the presence of metabolic inhibitors the initial rapid uptake by the nerve cord was reduced (Fig. 3). The high extracellular potassium level in this species is more difficult to explain, for it is not reduced in the presence of poison molecules and also appears to be independent of the concentration of labelled potassium ions in the slowly exchanging intracellular fraction. One explanation of the relatively high concentration of this cation might be that its activity coefficient is lower than that in free solution due to the presence of specific anion groups in the extracellular system.

The results summarized in Table II show that in poisoned preparations the distribution of sodium and of the two divalent cations is not inconsistent with that which would be expected in the event of a Donnan equilibrium with the haemolymph. It is difficult to relate the extracellular chloride concentration to that of the cations, due to the difficulties involved in defining the rapidly exchanging fraction for the anion in the nerve cord of *Carausius* (Treherne, 1964b).

It is now relevant to consider the effects of the composition of the extracellular fluid on the excitability of nerve cells of the insect central nervous system, and to relate this to the intracellular concentrations of the major inorganic ions. We have already seen that the ionic composition of the extracellular fluid in the cockroach central nervous system is equivalent to that of the body fluids of other animal groups, and that the high sodium concentration is achieved by a passive Donnan equilibrium with the haemolymph. The calculated extracellular and

TABLE III

The estimated cellular and extracellular ion concentrations in the nerve cord of *Carausius*. The values of sodium are based on *in vivo* studies, those for the remaining cations involve the extrapolation of *in vitro* results to the tissue concentrations shown in Table II (Treherne, 1964).

Cation	Concentration in haemolymph mmoles/l	Estimated concentration in extracellular fluid mmoles/l	Estimated intracellular concentration mmoles/l
Na	20·1	212·4	86·3
K	33·7	124·5	555·8
C	6·4	12·2	61·8
Mg	61·8	117·4	10·7

intracellular concentrations of the major inorganic ions in the central nervous tissues of *Carausius* are summarized in Table III. The potassium equilibrium potential calculated according to the Nernst equation from this data is approximately 37·1 mV. This must be regarded as a minimum estimate, for it has already been mentioned that the activity coefficient of the potassium in the extracellular phase may be reduced. This calculated equilibrium potential is low compared with the resting potentials which have been recorded in excitable cells of other insect species (cf. Narahashi, 1963), although the value of 37·1 mV for *Carausius* axons is close to the observed resting potential of 41·0 mV for the muscle fibres of this insect (Wood, 1957).

The sodium equilibrium potential for the central axons of *Carausius* calculated from the data in Table III is in the region of 22·3 mV, which is of the same order of magnitude as the 22·0 mV reversed potential recorded during the action potential of cockroach axons (Yamasaki and Narahashi, 1959). These considerations indicate, therefore, that despite the very specialized composition of the haemolymph the active secretion of sodium ions into the extracellular fluid from haemolymph produces an environment which is sufficient to account for the propagation of the action potential according to the classical membrane theory erected for the squid axon. This serves to explain the observation that removal of the very low concentration of sodium ions from the external medium resulted in the development of a rapid conduction block in the intact nerve cord of *Carausius* (Treherne, 1964a), for in the complete absence of external sodium the secretory mechanism would not be able to maintain the high extracellular concentration of the ion so that the concentration gradient with the axoplasm would be abolished.

The conduction processes in the nerve cord of *Carausius* have also been shown to be dependent upon the presence of external magnesium as well as sodium ions (Treherne, 1964a). Such an effect could result from some secondary effects of this divalent ion on the axon membranes. The calculated values summarized in Table III show, however, that the extracellular magnesium concentration exceeds the intracellular level of this cation, so that it is possible that, as with the muscle fibres of this insect (Wood, 1957), the action potential could result from an inward movement of the divalent cation. The equilibrium potential calculated from the magnesium concentrations shown in Table IV would, however, only amount to about 29·6 mV. The involvement of this divalent ion would thus only produce a slight increase in the mean potential as compared with that resulting from the movement of sodium ions alone, although it would contribute appreciably to the total ionic current carried during the action potential.

These results of the investigations outlined above show that the insect central nervous tissues are not virtually isolated beneath impermeable nerve sheaths, but are in a dynamic equilibrium with the smaller ions and molecules in the haemolymph. It can be further concluded that even in a phytophagous insect such as *Carausius* the composition of the fluid immediately in contact with the cells of the central nervous system approximates that found in the body fluids of most other animal groups and is sufficient to account for nerve function in terms of the membrane theory of conduction.

REFERENCES

Ashhurst, D. E. (1961). An acid mucopolysaccharide in cockroach ganglia. *Nature, Lond.* **191**, 1224–1225.

Hodgkin, A. L. (1951). The ionic basis of electrical activity in nerve and muscle. *Biol. Rev.* **26**, 339–409.

Hodgkin, A. L. (1958). Ionic movements and electrical activity in giant nerve fibres. *Proc. roy. Soc.* B. **148**, 1–37.

Hoyle, G. (1953). Potassium ions and insect nerve muscle. *J. exp. Biol.* **30**, 121–135.

Narahashi, T. (1963). The properties of insect axons. In "Advances in Insect Physiology" (J. W. L. Beament, J. E. Treherne, and V. B. Wigglesworth, eds.), Vol. 1, pp. 175–256. Academic Press, London and New York.

Shaw, J., and Stobbart, R. H. (1963). Osmotic and ionic regulation in insects. In "Advances in Insect Physiology" (J. W. L. Beament, J. E. Treherne and V. B. Wigglesworth, eds.), Vol. 1, pp. 315–399. Academic Press, London and New York.

Smith, D. S., and Treherne, J. E. (1963). Functional aspects of the organization of the insect nervous system. In "Advances in Insect Physiology" (J. W. L. Beament, J. E. Treherne and V. B. Wigglesworth, eds.), Vol. 1, pp. 401–484. Academic Press, London and New York.

Sutcliffe, D. W. (1963). The chemical composition of haemolymph in insects and some other arthropods in relation to their phylogeny. *Comp. Biochem. Physiol.* **9**, 121–135.

Treherne, J. E. (1961a). Sodium and potassium fluxes in the abdominal nerve cord of the cockroach, *Periplaneta americana*. *J. exp. Biol.* **38**, 315–322.

Treherne, J. E. (1961b). The movements of sodium ions in the isolated abdominal nerve cord of the cockroach, *Periplaneta americana*. *J. exp. Biol.* **38**, 629–636.

Treherne, J. E. (1961c). The efflux of sodium ions from the last abdominal ganglion of the cockroach, *Periplaneta americana*. *J. exp. Biol.* **38**, 729–736.

Treherne, J. E. (1961d). The kinetics of sodium transfer in the central nervous system of the cockroach, *Periplaneta americana*. *J. exp. Biol.* **38**, 737–746.

Treherne, J. E. (1962). The distribution and exchange of some ions and molecules in the central nervous system of *Periplaneta americana*. *J. exp. Biol.* **39**, 193–217.

Treherne, J. E. (1964a). Some preliminary observations on the effects of cations on conduction processes in the abdominal nerve cord of the stick insect, *Carausius morosus*. *J. exp. Biol.* (In press.)

Treherne, J. E. (1964b). The distribution and exchange of inorganic ions in the central nervous system of the stick insect, *Carausius morosus*. *J. exp. Biol.* (In press.)

Twarog, B. M., and Roeder, K. D. (1956). Properties of the connective tissue sheath of the cockroach abdominal nerve cord. *Biol. Bull., Woods Hole* **111**, 278–286.

Wood, D. W. (1957). The effects of ions upon neuromuscular transmission in a herbivorous insect. *J. Physiol.* **138**, 119–139.

Wyatt, G. R. (1961). The biochemistry of insect haemolymph. *Ann. Rev. Ent.* **65**, 75–102.

Yamasaki, T., and Narahashi, T. (1959). The effects of potassium and sodium ions on the resting and action potentials of the cockroach giant axon. *J. Insect Physiol.* **3**, 146–158.

The Free Amino Acid Pool of Cockroach (*Periplaneta americana*) Central Nervous System

J. W. RAY

Biochemistry Department, Agricultural Research Council, Pest Infestation Laboratory, Slough, Bucks, England

The biochemistry of insect nerve has received little attention apart from an extensive study of the acetylcholine system. The metabolism of insect nerve mitochondria is a relatively unexplored field. Flight muscle sarcosome preparations, on the other hand, have been extensively studied. In 1952 Sacktor and Bodenstein demonstrated the presence of cytochrome *c* oxidase in cockroach nerve and in 1955 Sacktor and Thomas demonstrated succino-cytochrome *c* reductase activity in the same tissue. The activity in both cases was lower than that found in flight muscle and hind gut. In 1958 Heslop and Ray reported the presence of ATP, ADP, AMP and glucose 6 phosphate in cockroach abdominal nerve. This study was extended, and in 1961 the same authors described the identification of the tri-, di- and mono-phosphates of cytosine, the tri- and di-phosphates of guanosine and uridine, DPN, TPN, glycerol 1-phosphate and arginine phosphate. Corrigan in 1959 identified proline, glutamate and glutamine in cockroach nerve. The presence of glycogen deposits in the cockroach nerve was demonstrated by Wigglesworth (1960), who showed that the deposits disappeared when the insect was starved and were reformed when the insect was fed. Treherne (1960) found that cockroach nerve took up [^{14}C]glucose and trehalose and metabolized them, liberating $^{14}CO_2$ and forming ^{14}C-labelled glutamate, glutamine, aspartate and alanine. Frontali (1962) has made a study of the glutamic acid decarboxylase activity of various subcellular fractions of bee brain, but he has not identified the morphological composition of these fractions.

The free amino acids of cockroach nerve were extracted and chromatographed as described by Ray (1964). Figure 1 shows a typical chromatogram of the free amino acids from nerve tissue. The amino acids have been visualized with ninhydrin. Some of the arginine will have been formed by the breakdown of arginine phosphate under the conditions of chromatography. Proline, glutamine and γ-amino-n-butyric

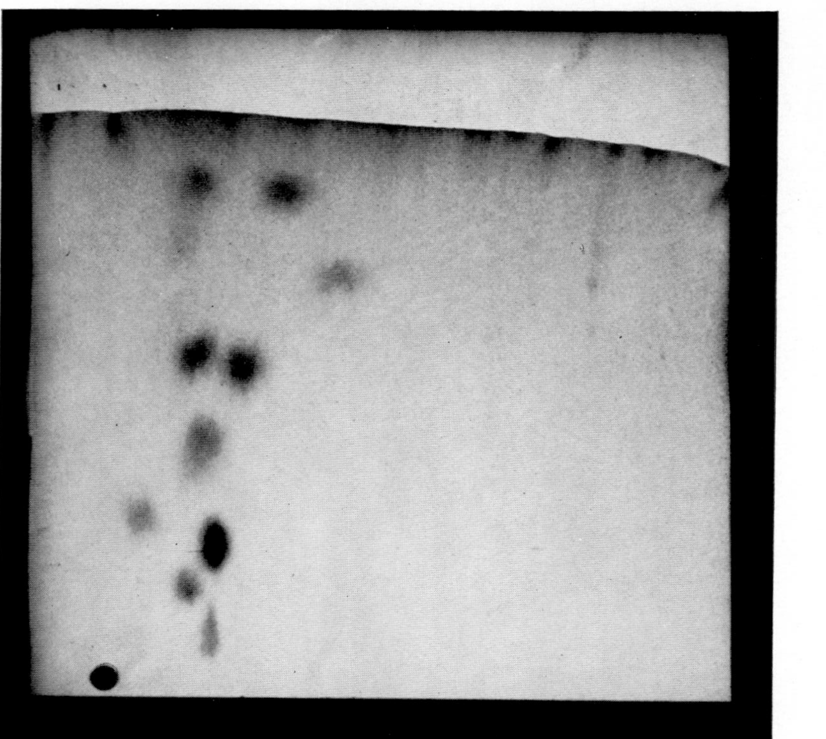

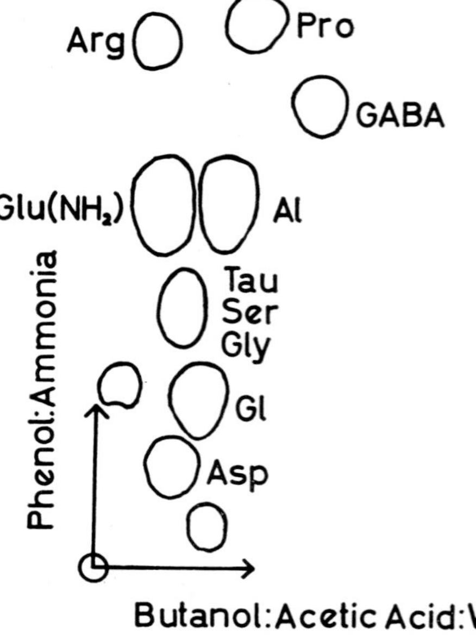

FIG. 1. Typical chromatogram of the free amino acids in the thoracic and abdominal nerve cord of *Periplaneta americana*. The amount of extract used is equivalent to one nerve cord.

Al, alanine; Asp, aspartate; GABA, γ-amino n-butyrate; Gl, glutamate; Glu(NH_2), glutamine; Pro, proline; Arg, arginine; Ser, serine; Gly, glycine; Tau, taurine. The two unlabelled spots are unknown.

acid are all metabolically linked with glutamate, which is derived from the Krebs cycle intermediate α-oxoglutarate. Aspartate and alanine are derivatives of the two keto acids, oxaloacetate and pyruvate respectively.

It was thought that the hyperactivity induced in the nervous system by many contact insecticides might provide a means for studying the metabolic role of the free amino acids. By comparing the changes in amino acid concentration following insecticide dosage with results obtained by injecting various inhibitory substances, it was hoped to gain further evidence for the metabolic pathways involved.

Table I shows the concentration of free amino acids in nerve from untreated cockroaches and cockroaches which have been prostrated by topical application of various insecticides. The results are expressed as μmoles/g wet weight of nerve. In all cases there has been a substantial depletion of the proline concentration in the nerve, even though the insecticides are chemically unrelated. DDT, DFP, the carbamate (*o*-isopropoxyphenyl-*N*-methyl carbamate) and dieldrin all cause a surge of respiration together with increased muscular and nervous activity. Bursell (1963) found that the high level of proline in the blood of the tsetse fly is rapidly depleted during flight. Flight also depresses the proline level in the thorax of the blowfly (Lewis and Fowler, personal communication). Thus, it would seem likely that the depletion of proline in these cases simply reflects the stimulated oxidation of this amino acid, some of the amino nitrogen appearing as glutamine as originally suggested by Winteringham (1958). However, n-valone is a depressant which lowers respiration (Gostick, 1960), muscular and nervous activity, and its effect on proline appears anomalous in this context. The action of n-valone has been investigated by Heslop and Ray (1964) on the aerobic and anaerobic phosphorus metabolism of the housefly thorax. The results showed that n-valone did not substantially deplete the concentration of high energy phosphates or appear to inhibit glycolysis in any way. Experiments with blowfly sarcosomes (Ray and Lewis, unpublished data) indicated that n-valone inhibited the oxidation of pyruvate and α-oxoglutarate. In n-valone poisoning proline is perhaps oxidized to satisfy energy requirements normally met by the Krebs cycle, even though the complete oxidation of proline is itself probably inhibited.

In the tsetse fly proline is quantitatively converted to alanine during flight (Bursell, 1963). Thus with DFP, carbamate and dieldrin poisoning, it is possible that the increase in alanine results from the oxidation of proline. During the hyperactive stage of DDT poisoning there was an increase in the alanine concentration, but this concentration fell at the

TABLE I

Effect of insecticides on free amino acids of cockroach thoracic and abdominal nerve cord expressed as μmoles/g wet weight

	Aspartate	Glutamate	Glutamine	Alanine	Proline	γ-Amino n-butyrate
Control	3·5 ± 0·9(6)	4·5 ± 0·25(6)	2·9 ± 0·6(6)	2·3 ± 0·5(6)	8·9 ± 1·7(6)	2·5 ± 0·8(6)
DDT	2·7	1·8	5·3	1·9	trace	1·8
DFP	4·2	3·3	5·8	8·0	1·7	3·2
Carbamate	2·8	2·2	4·1	7·4	1·7	3·5
Dieldrin	5·1	3·9	8·7	4·2	1·6	3·0
n-Valone	7·7	1·2	trace	27·6	trace	2·1

TABLE II

Effect of inhibitors on free amino acids of cockroach thoracic and abdominal nerve cord expressed as μmoles/g wet weight

	Aspartate	Glutamate	Glutamine	Alanine	Proline	γ-Amino n-butyrate
Control	3·5 ± 0·9(6)	4·5 ± 0·25(6)	2·9 ± 0·6(6)	2·3 ± 0·5(6)	8·9 ± 1·7(6)	2·5 ± 0·9(6)
Iodoacetate	2·5	2·0	5·0	2·2	0·9	4·0
Arsenite	2·1	4·3	4·5	12·9	3·9	5·0
Fluoroacetate	3·5	3·0	4·0	8·0	3·3	4·2
Rotenone	2·8	6·4	2·6	14·1	8·1	4·4

prostrate stage. Winteringham (1956), Winteringham *et al.* (1960) found that insects died of DDT poisoning without exhausting their endogenous reserves. It is possible that glycolysis is in some way inhibited during the later stages of DDT poisoning due to some nonspecific effect; for example, desiccation of the tissues interfering with the transport of metabolites. The alanine which had been accumulating now seems to be metabolized in lieu of pyruvate from glycolytic sources. During valone poisoning the accumulation of alanine is too large to have arisen solely from proline. It is also unlikely that the suggested pathway of proline oxidation can operate in the presence of valone, and therefore pyruvate is the most likely source of the alanine's carbon skeleton. There has been an increase in α-amino nitrogen in the fraction estimated, but its origin is unknown.

Table II shows the concentrations of free amino acids in the nerves of cockroaches prostrated with various inhibitors. Boccacci *et al.* (1960) showed that iodoacetate was able to penetrate the nerve of a cockroach and inhibit triose phosphate dehydrogenase. Iodoacetate has been shown to lower the aerobic resting potential of cockroach nerve (Yamasaki and Narahashi, 1957) indicating the importance of glycolysis in the energic metabolism of nerves. The results with iodoacetate almost exactly duplicate those found in the nerve of a DDT-prostrate cockroach, supporting the idea that a failure of glycolysis occurs at least in the late stage of DDT prostration.

With iodoacetate, arsenite and fluoroacetate there has been a substantial depletion of proline. This strengthens the arguments advanced previously that proline can and does act as a temporary metabolic reserve. Rotenone does not deplete the proline concentration, since it inhibits proline oxidation (Lewis and Fowler, personal communication). Arsenite inhibits pyruvate and α-oxoglutarate decarboxylation; n-valone has been shown to possess similar properties, and one might expect the amino acid changes to be similar in both cases. If the arsenite figures are compared with those of an earlier stage of n-valone poisoning, the similarity is striking except for the aspartate concentrations, which fall in arsenite poisoning and rise in n-valone poisoning.

As we have seen, an increase in the alanine concentration can arise in two ways, either as a result of the oxidation of proline during periods of stimulated respiration (e.g. during insecticide poisoning), or as an end product of glycolysis if pyruvate metabolism is blocked (e.g. in valone poisoning). We have two further examples illustrating this latter case. Both arsenite and rotenone inhibit pyruvate metabolism and bring about an increase in alanine concentration. A similar increase in the alanine concentration in nerve has been found under conditions of anoxia (Ray, 1964). It is interesting to compare this finding with the work of Price (1961), who found that, although glycolysis is known to continue for some time during anoxia in houseflies, there was only a transient increase in the pyruvate concentration, whilst the alanine concentration rose steadily, indicating that alanine can represent the end product of glycolysis in flies as well as in cockroach nerve.

The mode of action of fluoroacetate has been described by Peters (1957). Fluoroacetate takes part in a series of reactions in which it becomes activated by the acetate activating enzyme and attached to CoA to form fluoroacetyl CoA. Then in the presence of oxaloacetate and the condensing enzyme, it forms fluorocitrate, which inhibits aconitase and hence the Krebs cycle.

When the citrate concentration in the nerve of a fluoroacetate treated cockroach was estimated, it was found to be ten times greater than that in the control. The concentration of citrate in the haemolymph is higher than that in the nerve, and the increase in poisoned nerve could perhaps be explained simply by diffusion. However, a similar increase has been demonstrated in isolated nerve when it is incubated in a medium containing fluoroacetate. The nerve therefore contains the acetate activating enzymes, the condensing enzyme and probably aconitase.

Rotenone is the active constituent of the insecticide, derris. It has been shown by Ernster *et al.* (1963) to inhibit the oxidation of reduced DPN by the respiratory chain within the mitochondria. The lack of DPN within the mitochrondria would inhibit pyruvate metabolism. The accumulation of alanine indicates that glycolysis continues for some time and that the reduced DPN produced at the triose phosphate dehydrogenase step is oxidized via a hydrogen shuttle as described by Boxer and Devlin (1961). This shuttle is a system in which extra-mitochondrial-reduced DPN is oxidized by reducing a diffusable substance which then passes into the mitochondria and is oxidized. The substance then diffuses out, ready to start the cycle again. The most likely substrate is probably α-glycerol 1-phosphate, which has been shown to occur in nerve (Heslop and Ray, 1961).

Summary

1. The changes in the free amino acid pool of nerve induced by insecticides which cause hyperactivity are nonspecific and are a result of the hyperactivity.

2. Proline appears to act as a metabolic reserve in cases of hyperactivity

or on occasions when some part of the usual pathway of energy production is blocked.

3. Alanine can arise in two ways, either as a result of proline oxidation or as the end product of carbohydrate metabolism when pyruvate oxidation is blocked.

4. There is some evidence to support the idea that glycolysis fails in the late stages of DDT poisoning possibly due to some nonspecific effect.

5. Further evidence has been presented to support the idea that the Krebs cycle is active in insect nerve.

REFERENCES

Boccacci, M., Natalizi, G., and Bettini, S. (1960). Research on the mode of action of halogen containing thiol alkylating agents on insects. Effects of iodoacetate on choline acetylase. *J. Insect Physiol.* **4,** 20–26.

Boxer, G. E., and Devlin, T. M. (1961). Pathways of intracellular hydrogen transport. *Science* **134,** 1495–1501.

Bursell, E. (1963). Aspects of metabolism of amino acids in the tsetse fly Glossina (Diptera). *J. Insect Physiol.* **9,** 439–452.

Corrigan, J. J. (1959). "The metabolism of some free amino acids in DDT-poisoned *Periplaneta americana* L." Ph.D. Thesis, University of Illinois, Urbana.

Ernster, L., Dallner, G., and Azzoni, G. (1963). Differential effects of rotenone and amytal on mitochondrial electron and energy transfer. *J. biol. Chem.* **238,** 1124–1131.

Frontali, N. (1962). Comparative neurochemistry. *Proc. 5th Int. Neurochem. Symp.* 185–192.

Gostick, K. G. (1960). The relationships between increased oxygen uptake and locomotor ataxy or death in insecticide poisoned *Alphitobius laevigatus* F. *Ann. appl. Biol.* **49,** 46–54.

Heslop, J. P., and Ray, J. W. (1958). Phosphorus compounds of the cockroach nerve and effects of DDT. *Biochem. J.* **70,** 16.

Heslop, J. P., and Ray, J. W. (1961). Nucleotides and other phosphorus compounds of the cockroach central nervous system. *J. Insect Physiol.* **7,** 127–140.

Heslop, J. P., and Ray, J. W. (1964). Glycerol 1-phosphate metabolism in the housefly (*Musca domestica* L.) and the effects of poisons. *Biochem. J.* **91,** 187–195.

Peters, R. A. (1957). Mechanism of toxicity of the active constituents of *Dichapetalum cymosum* and related compounds. *Advanc. Enzymol.* **18,** 113–159.

Price, G. M. (1961). The accumulation of α-alanine in the housefly *Musca vicina*. *Biochem. J.* **81,** 15.

Ray, J. W. (1964). The free amino acid pool of the cockroach (*Periplaneta americana*) central nervous system and the effect of insecticides. *J. Insect Physiol.* **10,** 587–597.

Sacktor, B., and Bodenstein, D. (1952). Cytochrome c oxidase activity of various tissues of the American cockroach, *Periplaneta americana* L. *J. cell. comp. Physiol.* **40,** 157–161.

Sacktor, B., and Thomas, G. M. (1955). Succino-cytochrome c reductase activity of tissues of the American cockroach *Periplaneta americana* L. *J. cell. comp. Physiol.* **45**, 241–245.

Treherne, J. E. (1960). The nutrition of the central nervous system in the cockroach, *Periplaneta americana* L. The exchange and metabolism of sugars. *J. exp. Biol.* **37**, 513–533.

Wigglesworth, V. B. (1960). The nutrition of the central nervous system in the cockroach, *Periplaneta americana* L. The role of the perineurium and glial cells in the mobilization of reserves. *J. exp. Biol.* **37**, 500–512.

Winteringham, F. P. W. (1956). Resistance of insects to and acquired tolerance of insects to insecticides. *Chem. & Ind.*, 1182–1186.

Winteringham, F. P. W. (1958). Comparative aspects of insect biochemistry with particular reference to insecticidal action. *Proc. 4th Int. Congr. Biochem.* **12**, 201–210.

Winteringham, F. P. W., Hellyer, G. C., and McKay, M. A. (1960). Effects of the insecticides DDT and dieldrin on phosphorus metabolism of the adult housefly, *Musca domestica* L. *Biochem. J.* **76**, 543–548.

Yamasaki, T., and Narahashi, T. (1957). Effects of oxygen lack, metabolic inhibitors and DDT on the resting potential of insect nerve. Studies on the mechanism of action of insecticides. *Botyu-Kagaku* **22**, 259–276.

Synapses in the Insect Nervous System

D. S. SMITH

Department of Biology, University of Virginia, Charlottesville, Virginia, U.S.A.

The potential of the electron microscope as a tool in biological research first came to be realized a little over a decade ago through the introduction of suitable methods of fixation, embedding and thin-sectioning. Since that time, a great variety of animal tissues have been examined at high resolution and magnification, and these studies have afforded new insight into the structural basis of cell function and specialization. In the nervous system, which displays the most complex and intricate of all systems of cell relationships, many of the cytological features of greatest interest and importance are beyond the limit of resolution of the light microscope. The electron microscope is proving a valuable aid in the development of concepts relating recent advances in our understanding of the function of the elements of the central and peripheral nervous system to the details of fine structure of these cells.

In the nervous system of insects and vertebrates alike, discrete nerve cells—axons and their cell bodies—form an association in which sensory information and motor output are linked and integrated. The functioning of this very complex system depends on the passage of action potentials along the elongated axons and their branches and the transfer of excitation from one axon to the next, or from an axon to an effector cell, across the narrow intercellular discontinuities or "synaptic gaps" between successive cells. The purpose of this article is to review the available evidence concerning the cytological features of some of these synaptic areas in the insect nervous system, and to assess these findings in the light of corresponding physiological data. In this survey, the organization of two types of neuro-effector synapses will first be described, since these are morphologically the most easily defined, and later, interneuronal synapses in the central nervous system will be considered.

I. The Nerve-Muscle Junction

Every striated muscle cell or fibre in the animal body is intimately associated in one or more specialized regions with terminating motor axons from the central nervous system. At these neuromuscular junctions the arrival of an action potential at the nerve ending normally induces a similar depolarization of the adjoining muscle cell membrane, followed, after a short delay, by rapid contraction and relaxation of the fibre in a "twitch" or "fast" muscle, or by the enhancement of contraction in a "slow" muscle fibre. In vertebrate fast fibres, and in insect flight muscles other than those of Diptera, Hymenoptera, Coleoptera and certain Hemiptera, successive nerve action potentials elicit synchronized contractions or twitches in the innervated fibres. Whereas in vertebrates (e.g. frog) structurally distinct fibres showing either a slow tonic or a fast phasic response may occur together within a single anatomical muscle, in insect leg muscles no such structural differentiation of muscle cells is present, and the type of response elicited is determined synaptically through the presence of physiologically distinct slow and fast terminals at many of the neuromuscular junctions within the anatomical muscle (Hoyle, 1955). This last-mentioned double innervation does not, however, appear to occur either in synchronous or asynchronous insect flight muscles, and in most Orders the motor innervation of these fibres elicits only a fast phasic type of response (Pringle, 1957).

Light microscopic studies on these neuromuscular junctions have shown that the topography of the synaptic areas may vary considerably from one muscle to the next. The single motor end-plate of vertebrate fast fibres is a complex palmate structure; in synchronous insect muscles the nerve endings often meet the fibre surface in conical areas termed Doyère's cones, while in asynchronous flight muscles fine nerve branches pass longitudinally for some distance across the fibre surface. The electron microscope reveals that in all these instances, basic structural similarities between the nerve-muscle associations are present. The first of these features is that the cellular sheath that invests the peripheral axons along their course from the central nervous system is absent at least on one side of the terminating nerve branch, allowing the axon plasma membrane to lie very close to that of the underlying muscle fibre. A narrow gap is always retained between these two membranes, this gap being about 400–500 Å units wide in vertebrate junctions, or as little as 100 Å units wide in those of insect muscles. The signal passes from nerve to muscle across this "synaptic gap", and it is known that in vertebrate skeletal muscle this signal consists of the secretion of acetylcholine from the nerve. This substance alters the ionic permeability properties of the muscle membrane beneath the junction, initiating a local depolarization or "end-plate potential", which in turn gives rise to the propagated muscle action potential.

Palade (1954) first observed that the cytoplasm of the axon terminal presents a very characteristic appearance in the electron microscope, containing many small vesicles, and numerous mitochondria. These vesicles were subsequently found by many investigators in a variety of vertebrate neuromuscular junctions, and morphologically similar structures were likewise found in insect junctions on synchronous and asynchronous fibres by Edwards (1959, 1960), Edwards and co-workers (1958a, b), Smith (1960) and Smith and Treherne (1963). The last reference includes a review of the cytology of these junctional areas in different muscles. Figure 1 illustrates the organization of a neuromuscular junction on a leg muscle fibre of the honey-bee *Apis*, and large numbers of the "synaptic vesicles" are seen in the terminal axoplasm. It has been estimated that these vesicles are present in a concentration of *ca* 5000–7500 per cubic micron in this junction and in endings on asynchronous flight muscle fibres of the beetle *Tenebrio* (Smith, 1960). In vertebrates and insects alike, the synaptic vesicles are *ca* 250–450 Å in diameter.

Katz, Fatt and del Castillo have united the electron microscopic observations on the vertebrate neuromuscular junction with physiological and pharmacological data, producing an attractive hypothesis of the mechanism of synaptic transmission (refs. in Katz, 1962). They found that in the resting muscle fibre spontaneous electrical activity, consisting of very small-amplitude (0·5 mV) depolarizations, occurs in the end-plate region, and that these "miniature end-plate potentials" represent the effect on the post-synaptic (muscle) membrane of minute quantities of the chemical transmitter, acetylcholine, liberated from the axon terminal. They suggested that the transmitter is encapsuled within the synaptic vesicles in the terminal axoplasm, and that the small "quantal" depolarizations occur when the vesicles collide with "reactive sites" on the axon membrane, discharging their contents into the synaptic gap and thence to the post-synaptic membrane. Extending this hypothesis, they suggested that when the nerve action potential arrives at the terminal, the frequency of this occurrence is momentarily enormously increased, with several hundred "packets" of acetylcholine being released in less than one millisecond and combining to elicit the full end-plate potential which, when it passes a threshhold value, triggers the propagated muscle action potential. The transmitter is believed to be rapidly destroyed by the acetylcholinesterase associated with the postsynaptic membrane, and the mechanism is thus reset to receive the next impulse in the motor nerve train.

Can the above hypothesis be employed to cover the case of the neuromuscular junctions of insects? The morphology of the synapse certainly

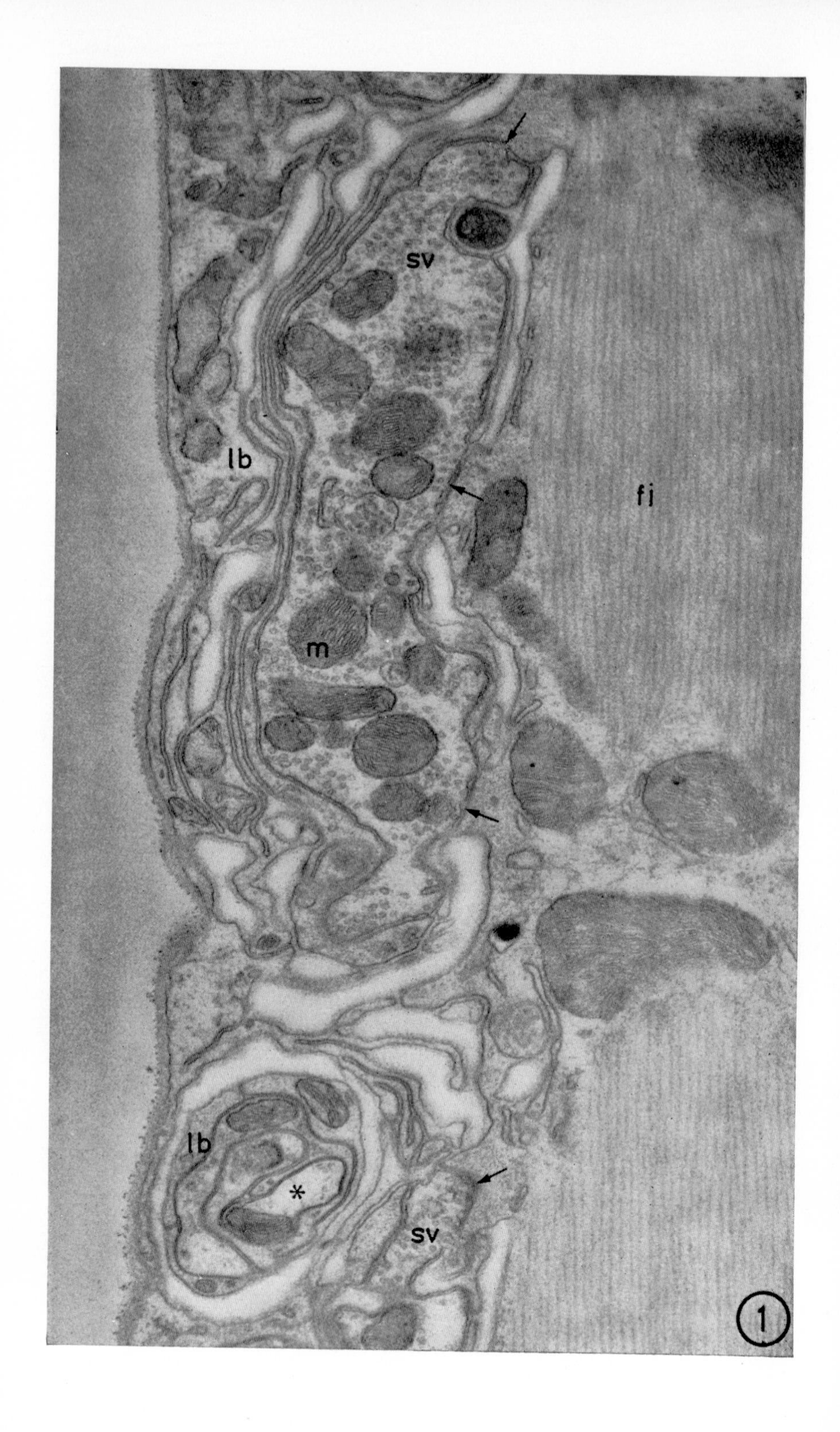
sv
lb
fj
m
lb
*
sv
1

suggests a functional similarity between vertebrate and insect junctions, but it is clear that acetylcholine is not the transmitter in the latter. Indeed, the substance that performs this function in insects has yet to be identified. However, it is known that the first membrane response after the arrival of a nerve impulse at an insect junction is, as in vertebrate muscle, an end-plate potential. Moreover, Usherwood (1961) has identified spontaneous miniature end-plate potentials in preparations of resting leg muscles of *Schistocerca* and *Blaberus* which are similar in amplitude and frequency to those of vertebrate endings. It therefore seems probable that the insect and vertebrate neuromuscular junctions employ the same mechanism for making the transmitter available to the postsynaptic membrane, whilst differing in the chemical nature of the transmitter molecule.

No structural features permitting the identification of an axon as "fast" or "slow" have been found in insect neuromuscular junctions containing such double innervation, but it may well be that the basis of this physiological differentiation resides in the presence of additional transmitter substances, unaccompanied by any distinguishing cytological variation.

Although the hypothesis outlined above is in good accord with the cytological and physiological data, it must be realized that no information is at present available either in insects or vertebrates on the site of synthesis of the transmitter or on the mode of formation of the synaptic vesicles. Possibly, in each instance these structures are formed within the terminal axoplasm; electron microscopic studies give no indication that they are elaborated proximally in the axon or cell body of the neurone.

II. The Firefly Luminescent Organ

The luminescent organs or "lanterns" of adult fireflies afford a second example of a nerve-effector system, since the production of light *in vivo* is under the control of the central nervous system *via* a peripheral motor nerve supply. In adults of the North American species of lampyrids

←

Fig. 1. Electron micrograph of a neuromuscular junction in honeybee (*Apis*) coxal muscle. The contractile fibrils (fi) of the fibre occupy the right-hand part of the field. Terminating axons contain large numbers of "synaptic vesicles" (sv) and mitochondria (m), and localized regions where muscle and nerve plasma membranes are in close apposition are indicated by arrows. In this type of muscle, the synapsing nerve terminals are capped externally by the Schwann cell or lemnoblast (lb): the small nerve branch at lower left (asterisk) contains no synaptic vesicles and is still surrounded by the complete lemnoblast sheath characteristic of nerves proximal to the region of synapse. (From Smith and Treherne, 1963.) × 31,000.

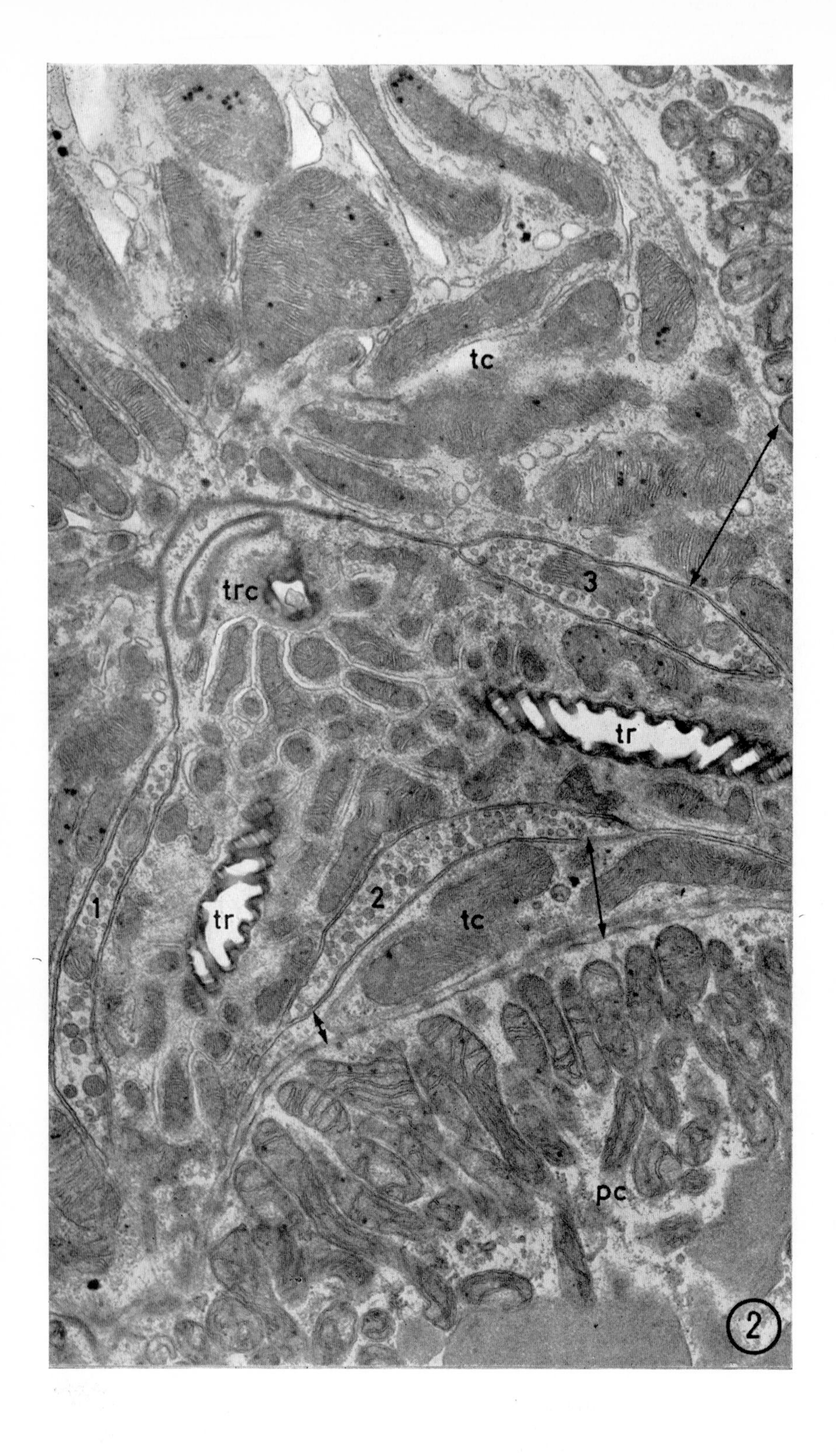

tc
trc
3
tr
tr
1
2
tc
pc
2

(e.g. of *Photuris* and *Photinus*), light is emitted in brief flashes from the lanterns situated on the ventral surface of one or two of the abdominal segments, while in species of the tropical elaterid genus *Pyrophorus*, in which the chief luminescent organs lie in the prothorax, light emission may last for several seconds. The former instance has been more extensively studied than the latter from both the cytological and physiological standpoint, and has been selected for discussion here.

The firefly lantern is a structure of great complexity, and details of the organization of all the cells included in its makeup need not be given at this point: an electron microscopic account of these organs in *Photuris pennsylvanica* has been published by Smith (1963). It may be said, in summary, that the bulk of each lantern consists of highly specialized cells, the photocytes, in which the light is believed to be produced. Buck (1948), in a valuable review of the histology and physiology of insect luminescent organs, states that in males of *Photinus pyralis* the two lanterns together comprise about 15,000 photocytes. These photocytes are arranged in a precise pattern around about 600 cylindrical channels which traverse the lanterns. Into these channels pass tracheal trunks, giving rise to laterally oriented tracheal end-cells, 80–100 of which lie in each cylinder. These end-cells and the tracheoles that extend beyond them are in very close association with the photocytes. Into the cylinders, together with the tracheae, pass branches of motor nerves from the abdominal ganglia. Hanson (1962) observed that the physiological unit of flashing is not the whole lantern but the group of cylinders and associated photocytes supplied by the finest peripheral nerve branches.

A considerable amount is known of the *in vitro* biochemistry of the firefly luciferin-luciferase system responsible for light production (McElroy and Hastings, 1955; McElroy and Seliger, 1961), and, moreover, many of the neurophysiological properties of the intact system have been determined (Case and Buck, 1959; Hanson, 1962). However, light microscopic studies failed to demonstrate the distribution of nerve terminals within the lantern; a consequence of the complexity of the organ and the small size of the terminating nerve branches. The high resolving power of the electron microscope added this detail to our

←

FIG. 2. Electron micrograph of a tracheal cell–nerve ending complex in the luminescent organ of the firefly *Photuris pennsylvanica*. The terminal processes of the motor nerve (1, 2, 3) are tightly inserted between the cell membrane of the tracheolar cell (trc) and that of the surrounding tracheal end-cell (tc), and the axoplasm of the synaptic region contains mitochondria and large numbers of vesicles. The nerve terminals are separated from the surface of the effector cells, the photocytes (pc), by end-cell membranes and cytoplasm (between arrows). Note the tracheolar branches (tr). (From Smith, 1963.) × 23,000.

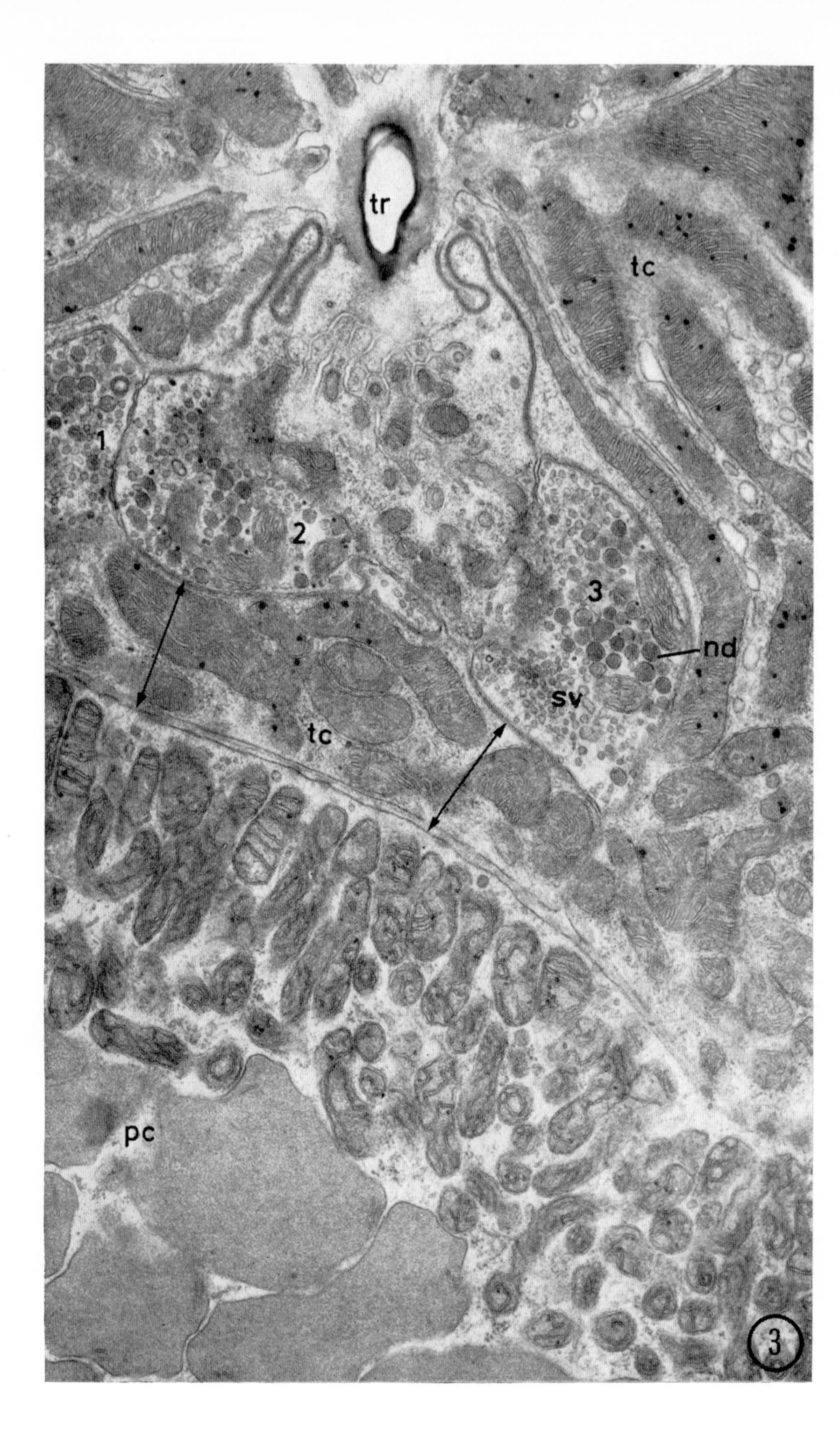

tr
tc
1
2
3
nd
sv
tc
pc
3

knowledge of the lantern, and the result obtained was in some ways surprising. The nerve terminals were found to lie not directly against the effector cells (the photocytes), but between the surfaces of the tracheal end-cell and tracheolar cell, that is, up to about one micron from the edge of the adjoining photocytes. Figure 2 represents an electron micrograph of this end-cell–tracheolar cell association and includes profiles of three synapsing terminal axon branches. Within the latter are found two structures also present in the axoplasm at the neuromuscular junction: numerous mitochondria and large numbers of vesicles. Unlike the situation in muscle, however, these vesicles are dimorphic in the lantern ending; one population is *ca* 250–300 Å in diameter, resembling the synaptic vesicles of the muscle junction, and these are interspersed with much larger vesicles *ca* 1000–1500 Å in diameter, similar to the "neurosecretory droplets" that have been found in axons within the central nervous system and elsewhere. These two axoplasmic components are more clearly seen in sections that pass tangentially through the broad axon terminals, as in Fig. 3.

If the presence of vesicular structures both in the neuromuscular junction and in the firefly lantern indicates that they share a common mechanism of transmitter release, then we are faced with the dual problem of determining the nature of the material or materials secreted from the terminals in the lantern and of explaining the fact that the immediate recipients of any such secretions appear to be the tracheal cells surrounding the synapse, and not the photocytes, which are considered to be the effector cells. One striking physiological difference between the lantern synapse and the neuromuscular junction is that the delay between stimulation of the nerves supplying the lantern and the resulting flash, about 65–75 msec, is far longer than the corresponding delay in a nerve-muscle preparation. The morphology of the lantern synapse is not inconsistent with this peculiarity if it is assumed that the photocytes are affected secondarily *via* events taking place in the tracheal cytoplasm interposed between them and the nerve terminals, whereas in the neuromuscular junction release of the transmitter from the presynaptic membrane places this molecule in an extracellular space within 100–500 Å of the effector cell membrane.

←

FIG. 3. A field similar to that shown in Fig. 2, but in which the terminal nerve processes are seen in tangential section. The axoplasm of these processes (1, 2, 3) contains mitochondria, small vesicles (sv) resembling the synaptic vesicles of other junctions, and larger vesicles with a dense content (nd) resembling neurosecretory droplets. The lower part of the field is occupied by the edge of a photocyte (pc), and a portion of the tracheal end-cell (tc) is interposed between this photocyte and the nerve terminals. A tracheole is seen at tr. (From Smith, 1963.) × 26,000.

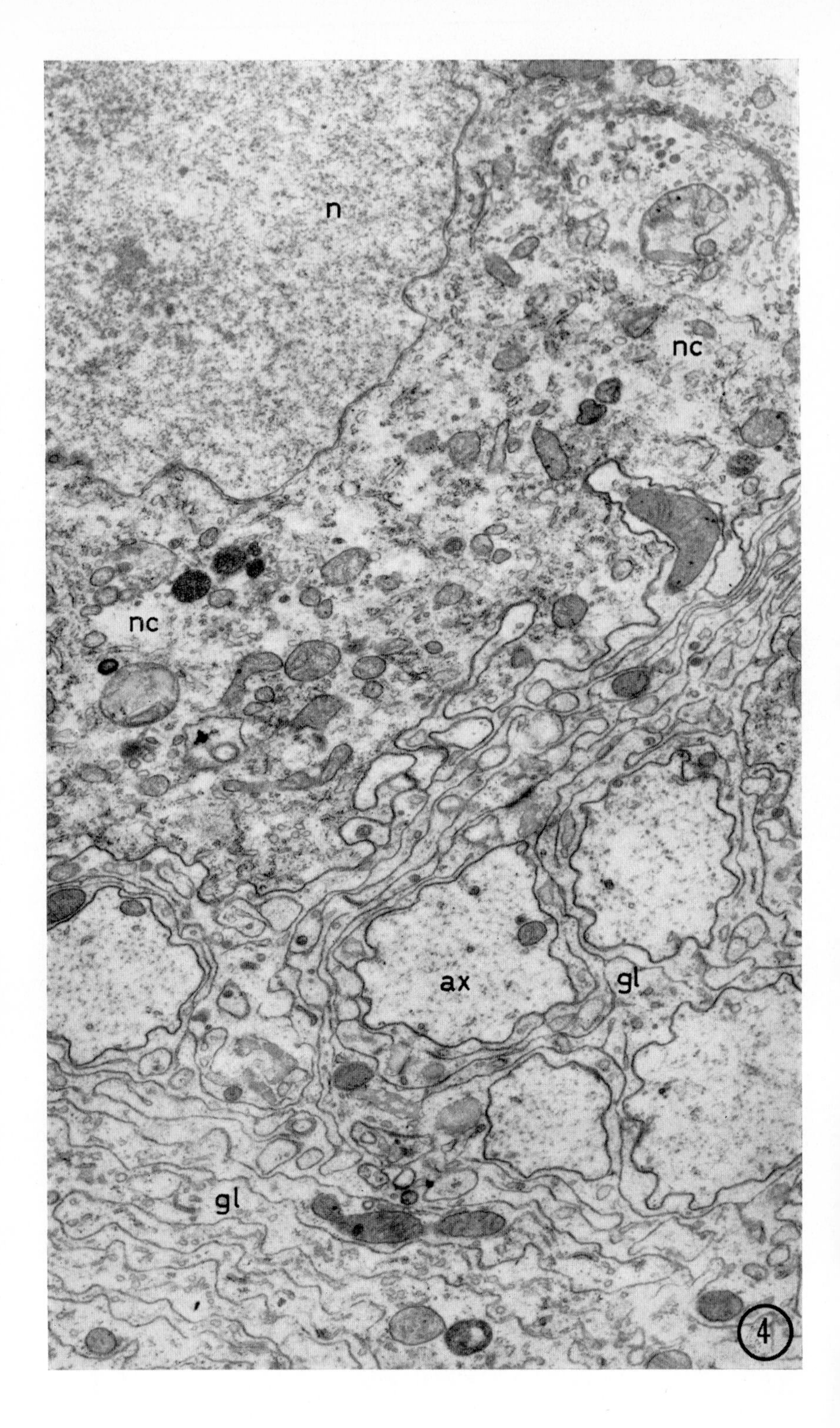
n
nc
nc
ax
gl
gl
4

Beyond this point, we can only speculate at present on the nature of the synaptic activation of the photocytes. The *in vitro* experiments of McElroy and Hastings (1955) and McElroy and Seliger (1961) suggest that the trigger for light production in the photocytes may be the arrival within them of a "pulse" of pyrophosphate rather than membrane depolarization. These investigators outline a hypothetical scheme involving the liberation of acetylcholine from the nerve terminals, resulting in the production of pyrophosphate *via* an acetylcholine–Coenzyme A–adenosine triphosphate cycle. It is possible that the biochemical mechanism controlling this cycle resides in the tracheal end-cell, which is very richly supplied with mitochondria (Figs. 2, 3), and that the long delay between stimulation and flashing reflects the production of the triggering pyrophosphate in the end-cell and its transfer to the adjoining photocytes. However, it must be stressed that acetylcholine has not yet been identified in this synapse, and that in any case the scheme outlined above may well be an oversimplification, since the presence of two distinct types of vesicle in the axoplasm of the nerve ending may prove to be connected with a more complex system of transmitter release.

We have already seen from a comparison of the neuromuscular junction in vertebrates and insects that different transmitters can perhaps be supplied by a common mechanism. Despite the present uncertainty concerning the biochemical details of *in vivo* light production in the firefly, the cytological similarity of the axoplasm in all these nerve endings is striking, and may reflect a basically similar mechanism of transmitter release in muscle and the luminescent organ, masked by differences in the neuro-effector architecture and by the dissimilar effector responses.

III. The Central Nervous System

The ganglia of the insect central nervous system are surrounded by an envelope of extracellular material containing collagen-like fibrils, the

←

Fig. 4. Electron micrograph illustrating the relationship between a nerve cell body and glial and axonal components at the surface of the neuropile in the last abdominal ganglion of the cockroach, *Periplaneta americana*. The upper portion of the field is occupied by the nucleus (n) and cytoplasm (nc) of a neurone, and five transversely sectioned axons (ax) are present in the lower part of the micrograph. The axons are surrounded by a layered system of glial processes (gl), and the latter also surround and indent the surface of the nerve cell body. × 20,000.

N.B. The material illustrated in this micrograph and in Figs. 5, 6 and 7 was fixed in glutaraldehyde and postosmicated. All other figures are of osmium tetroxide-fixed material.

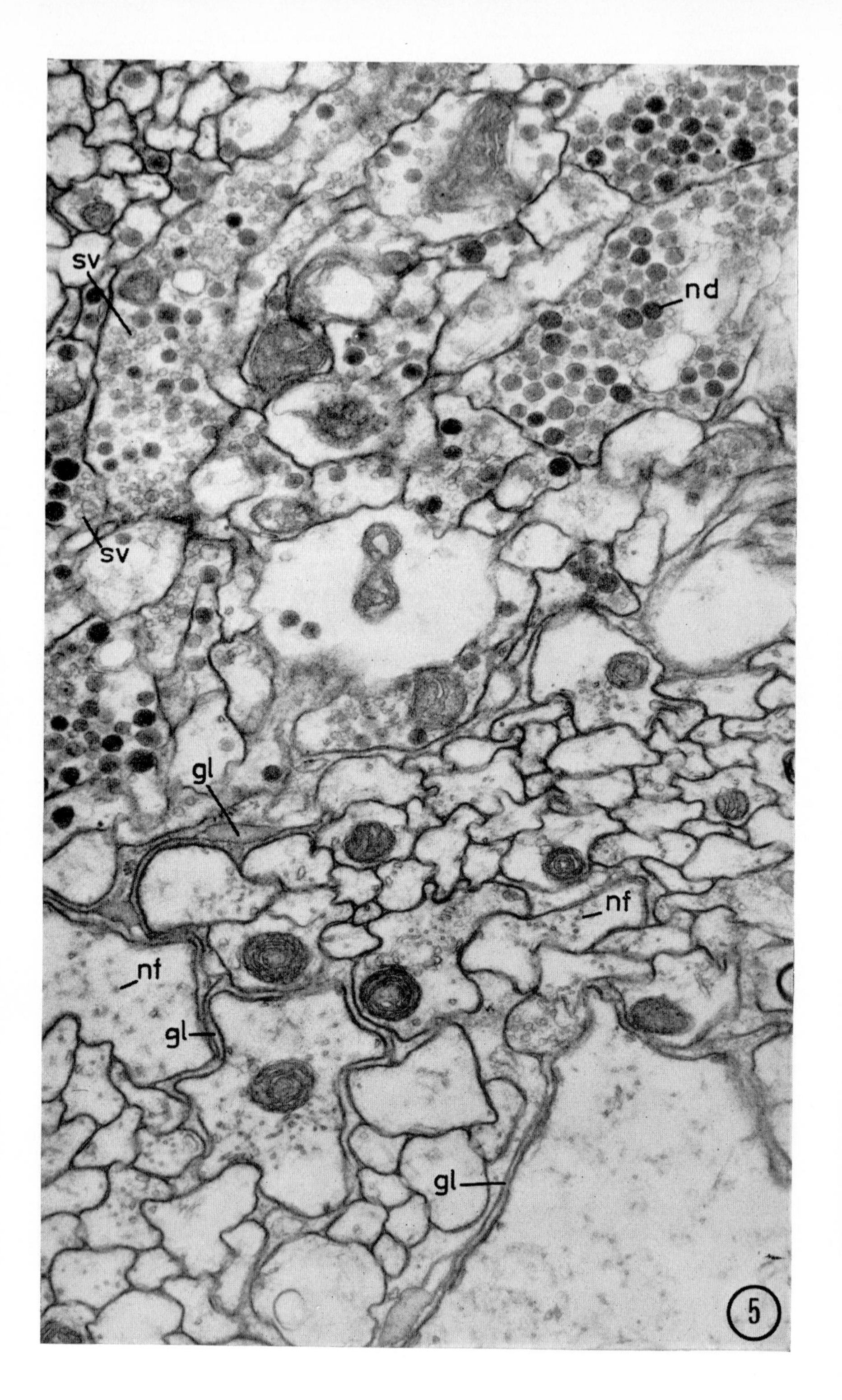
sv
nd
sv
gl
nf
nf
gl
gl
5

neural lamella, overlying a cellular layer, the perineurium, rich in mitochondria and glycogen deposits. The cell bodies or perikarya of the neurones are situated peripherally in the ganglion beneath the perineurium, and they are interspersed with and encapsulated by complex cytoplasmic processes of the glial cells (Fig. 4). Whereas in the vertebrate central nervous system synapses are often effected between axon terminals and the neurone cell body or its dendritic prolongations, the glial insulation does not permit such axo-somatic or axo-dendritic synapses in the insect ganglion, and the axo-axonic synaptic pathways of the insect central nervous system, involving sensory, motor and internunciary components, are relegated to the central region of the ganglion—the neuropile. A fuller account of the above structures in the last abdominal ganglion of *Periplaneta* and of those to be described later may be found in the review by Smith and Treherne (1963).

Electron micrographs of the neuropile show, at a different level of magnification and resolution, a similar picture to that provided by the light microscope: a highly complex system of profiles of axons and axon branches, ranging in diameter from about 0·15 micron to several microns. Within this interlacing system the conduction and synaptic pathways are established, and our present concern is to determine whether the cytological structure of the cell processes in the neuropile gives any clue to the distribution of the synaptic links within it. In an electron micrograph of any region of the neuropile, it is clear that in addition to the axon profiles many narrow glial processes are present. A typical field showing this organization is reproduced in Fig. 5. These glial prolongations originate in the gliocyte cell bodies situated at the surface of the neuropile, or occasionally within the neuropile itself, and they are situated in such a way that adjacent axon profiles may either be separated by them or, where they are absent, allowed to come into very close apposition. By analogy with neuro-effector junctions, it seems likely that synapses within the central nervous system require such close apposition of pre- and postsynaptic elements, and thus the glial processes within the neuropile may act as insulators, preventing or modifying chemical or electrical transmission when interposed between axons, and in this way delimit the pattern of excitation transfer within the ganglion. For further indications of the distribution of synapses

←

FIG. 5. Electron micrograph of a portion of the neuropile in the last abdominal ganglion of *Periplaneta*, including many transversely sectioned axon profiles. In some regions, narrow glial processes (gl) are present between the axons. The axoplasmic components comprise mitochondria, neurofilaments (nf), small synaptic vesicles (sv) and larger neurosecretory droplets (nd). × 33,000.

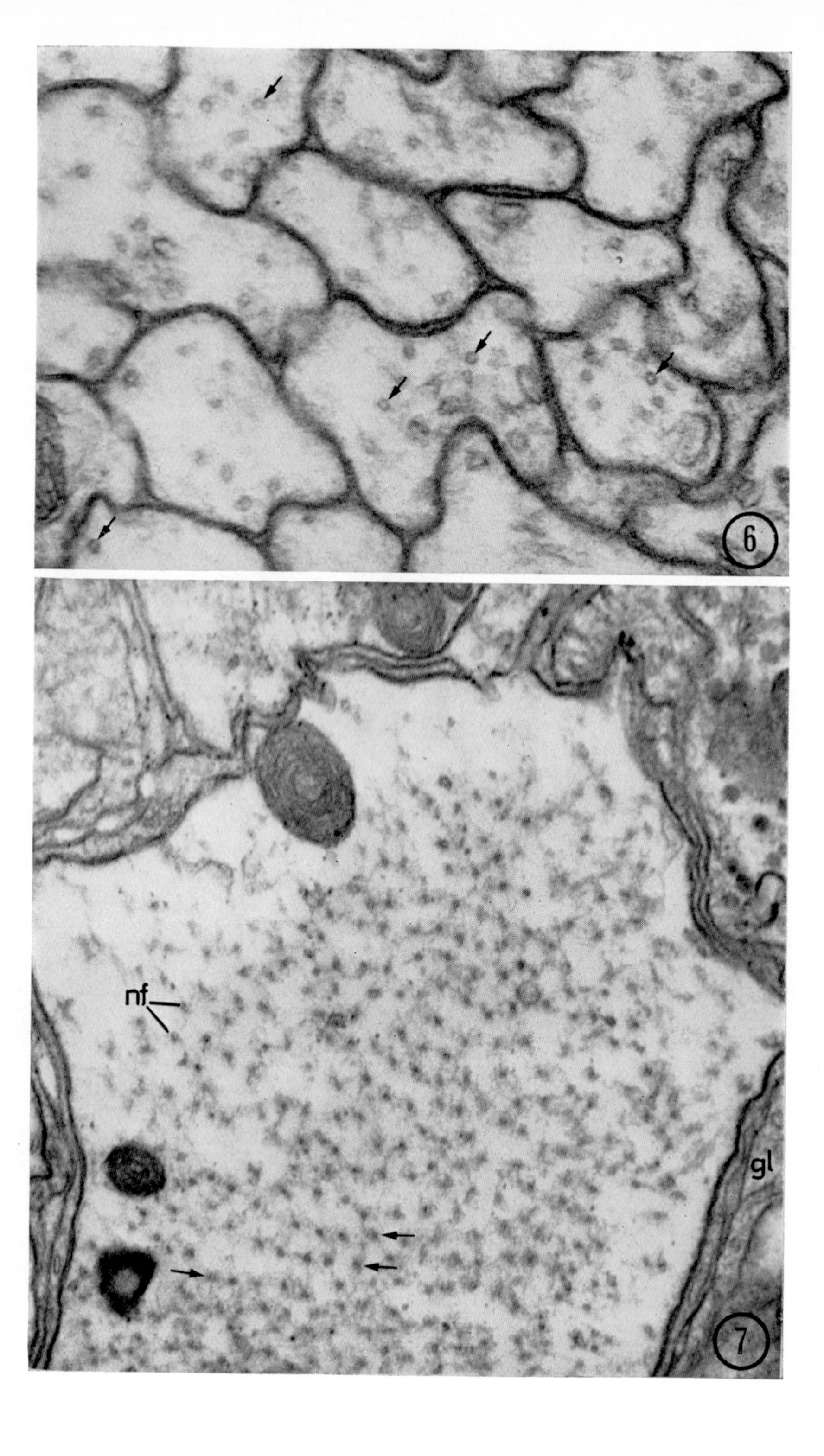
6
nf
gl
7

in this part of the nervous system, we must examine the cytological structure of its components in more detail.

The axon profiles within the neuropile contain a number of structurally well-defined units: mitochondria, neurofilaments and membrane-limited vesicles. The neurofilaments are seen in transverse section (Fig. 6) as circular profiles, *ca* 100–150 Å in diameter, including an electron-transparent "core". In some axons (Fig. 7) a suggestion of non-random arrangement of these filaments is seen. Longitudinal sections of axons indicate that these structures are of considerable though undetermined length. Similar components have been described in the axoplasm of vertebrates and other animals, and the composition and appearance of the units of squid giant fibres has been investigated in detail by Schmitt and Davison (1961). These authors found that squid neurofilaments consist of a protein, and on the basis of the observed dissociation or "unravelling" of the filaments outside the pH range of 6–8, they proposed that each intact filament may be built up of linear aggregates of small protein subunits, one or more of which may be linked together in a helical arrangement in the complete structure. The function of the neurofilaments is not at present known, but as Schmitt and Davison point out, it seems probable that such a ubiquitous axoplasmic structure serves some important function in the physiology of the axon. The neurofilaments are not, however, a characteristic component of neuro-effector terminals or of synaptic areas within the central nerve chain of insects or other animals.

While many axon profiles wihin the insect neuropile contain only neurofilaments and small numbers of mitochondria, others include, in varying proportions, small membrane-limited vesicles with a light content, and larger vesicles each having an electron-dense content. The small vesicles resemble the "synaptic vesicles" of the neuromuscular junction, and the larger ones represent the "neurosecretory droplets". As in the case of the terminal axoplasm at neuro-effector junctions, in the neuropile mitochondria are most numerous in axon profiles containing these vesicular structures. In the vertebrate central nervous system, in situations where the pre- and postsynaptic members of a synapse

←

FIG. 6. Higher power electron micrograph of a group of small axon profiles in the neuropile of *Periplaneta*. Numerous transversely sectioned neurofilaments are included, and in many instances (arrows) the electron-transparent central region of these neurofilaments is resolved. × 110,000.

FIG. 7. A transversely sectioned axon within the neuropile of the last abdominal ganglion of *Periplaneta*. The axoplasm contains large numbers of neurofilaments (nf), and some evidence of an ordered arrangement of these is seen in the regions indicated with arrows, and elsewhere. This axon profile is surrounded by a glial sheath (gl). × 60,000.

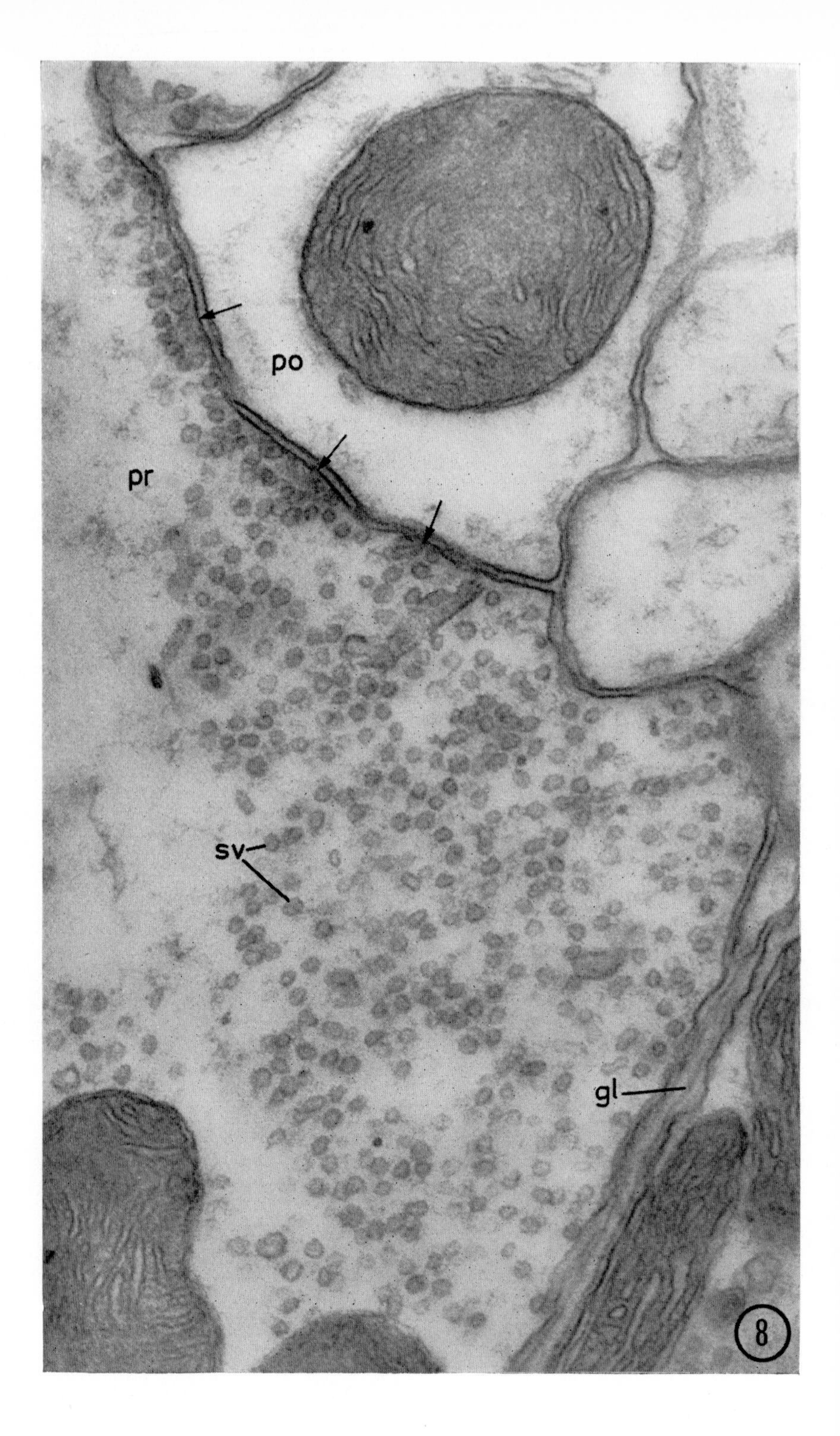
po
pr
sv
gl
8

could be identified with certainty, De Robertis (1955, 1956) and Palay (1958) found that the small synaptic vesicles were not distributed randomly in the presynaptic axoplasm, but that many of these were aggregated into clusters about 1500–5000 Å in width, adjoining the cell membrane of the presynaptic axon. Similar aggregations have been described in the vertebrate neuromuscular junction (Katz, 1962). Following the hypothesis that the synaptic vesicles contain the chemical transmitter discharged into the extracellular gap between the two elements of the synapse on arrival of the action potential in the presynaptic terminal, Palay (1958) proposed that ". . . the complex of a cluster of synaptic vesicles, associated with a focalized area of presynaptic plasmalemma, and the synaptic cleft may be considered as a morphological subunit" of the synapse, representing the specialized region at which transmitter release and excitation transfer takes place.

At present, we have no means of identifying the origin of a particular axon profile within the neuropile. However, the characteristic configuration of membranes and vesicles described above occurs within this region of the insect ganglion, as is shown in Fig. 8. This electron micrograph shows a region of close apposition between two axon profiles, lacking any interposed glial material: in one profile, vesicular structures are virtually absent, while in the other these structures are abundant, and show the same "focal" arrangement that was noted in vertebrate synapses. A further similarity to the latter is found in the area of dense axoplasm within the presumed postsynaptic member, opposite each presynaptic vesicular cluster (Fig. 8).

Trujillo-Cenóz (1959, 1962) pointed out that close proximity between axon branches, and the presence of vesicles within the axoplasm, cannot alone be indicative of synaptic sites, since these features are of too general occurrence in the neuropile. It now seems likely that chemically triggered excitation may take place across very localized regions between closely adjoining axons, where glial insulation is absent, and that the synaptic foci recognized in the neuropile may represent specialized

←

FIG. 8. Electron micrograph of a region in the neuropile of *Periplaneta*, including an area believed to represent a synaptic junction between presynaptic (pr) and postsynaptic (po) axons. The axoplasm of the former contains large numbers of synaptic vesicles, which are absent from the latter. Along the plasma membrane of the presynaptic component, the vesicles are collected into clusters (arrows), and each cluster or focus is backed in the postsynaptic axoplasm by a localized region of higher density. These "synaptic foci" were first identified in the central nervous system of vertebrates, by De Robertis (1955, 1956) and Palay (1958). Note that glial processes present elsewhere (gl) are absent from the synaptic region, permitting very close apposition of pre- and postsynaptic cell membranes. (From Smith and Treherne, 1963.) × 63,000.

areas in which synaptic vesicles are able to discharge transmitter molecules from the presynaptic membrane close to correspondingly localized receptive sites on the membrane of the postsynaptic element.

In conclusion, it must not be forgotten that the electron microscope at the present time probably shows us only a small part of the whole picture, or, rather, that we are able to interpret only to a limited extent the functional significance of the structures that we see in electron micrographs. A comparison between vertebrate and insect nerve-muscle junctions has shown that similar cytological organization may accompany chemically different synapses, and the same is probably true of synapses in the insect central nervous system. Although acetylcholine and acetylcholinesterase are known to be present in insect ganglia, the distribution of these substances has not yet been localized cytochemically, and the extent of the role of acetylcholine or other transmitters in central synaptic function is as yet undetermined. Furthermore, we have no information on the physiological role of the "neurosecretory droplets" within the ganglion, but perhaps here, as may be the case in the firefly luminescent organ, certain chemical synapses are controlled by more than one transmitter substance.

Note Added in Proof

Since this chapter went to press, cytochemical information has been obtained on the localization of sites of cholinesterase activity within the last abdominal ganglion of the cockroach, *Periplaneta* (Smith and Treherne, in preparation). The method used by R. J. Barnett (*J. Cell Biol.*, 1962, **12,** 247) in a study of the vertebrate neuromuscular junction was employed to assess the distribution of this enzyme, at the electron microscopic level. Of particular interest here is the observation that much of the eserine-sensitive esterase (cholinesterase) within the neuropile is distributed in discontinuous patches, often adjoined by clusters of synaptic vesicles; that is, in the vicinity of the "synaptic foci" mentioned previously. This evidence points to the existence of a cholinergic mechanism controlling at least some of the regions of synaptic contact within the neuropile of this insect.

REFERENCES

Buck, J. B. (1948). The anatomy and physiology of the light organ in fireflies. *Ann. N. Y. Acad. Sci.* **49,** 397–482.

Case, J. F., and Buck, J. B. (1959). Peripheral aspects of firefly excitation. *Biol. Bull.* **117,** 407.

De Robertis, E. (1955). Changes in the "synaptic vesicles" of the ventral acoustic ganglion after nerve section. *Anat. Rec.* **121,** 284–285.

De Robertis, E. (1956). Submicroscopic changes of the synapse after nerve section in the acoustic ganglion of the guinea pig. *J. biophys. biochem. Cytol.* **2,** 503–512.

Edwards, G. A. (1959). The fine structure of a multiterminal innervation of an insect muscle. *J. biophys. biochem. Cytol.* **5**, 241–244.

Edwards, G. A. (1960). Comparative studies on the fine structure of motor units. *In* "Vierter Int. Kongr. Elektronmikroskopie, 1958", Vol. 2, pp. 301–308. Springer, Berlin.

Edwards, G. A., Ruska, H., and Harven, E. de (1958a). Electron microscopy of peripheral nerves and neuromuscular junctions in the wasp leg. *J. biophys. biochem. Cytol.* **4**, 107–114.

Edwards, G. A., Ruska, H., and Harven, E. de (1958b). Neuromuscular junctions in the flight and tymbal muscles of the Cicada. *J. biophys. biochem. Cytol.* **4**, 251–256.

Hanson, F. E. (1962). Observation on the gross innervation of the firefly light organ. *J. Insect Physiol.* **8**, 105–111.

Hoyle, G. (1955). Neuromuscular mechanisms of a locust skeletal muscle. *Proc. roy. Soc.* B. **143**, 343–367.

Katz, B. (1962). The Croonian Lecture: The transmission of impulses from nerve to muscle and the subcellular unit of synaptic action. *Proc. roy. Soc.* B. **155**, 455–477.

McElroy, W. D., and Hastings, J. W. (1955). *In* "The Luminescence of Biological Systems" (F. H. Johnson, ed.), p. 161, A.A.A.S. Publications, Washington, D.C.

McElroy, W. D., and Seliger, H. H. (1961). *In* "Light and Life" (W. D. McElroy and B. Glass, eds.), p. 219. Johns Hopkins Press, Baltimore.

Palade, G. E. (1954). Electron microscope observation of interneuronal and neuromuscular synapses. *Anat. Rec.* **118**, 335–336.

Palay, S. L. (1958). The morphology of synapses in the central nervous system. *Exp. Cell Res.*, suppl. **5**, 275–293.

Pringle, J. W. S. (1957). "Insect Flight." Cambridge University Press.

Schmitt, F. O., and Davison, P. F. (1961). Biologie moléculaire des neuro-filaments. *In* "Actualités Neurophysiologiques" 3ème sér. (A. M. Monnier, ed.), pp. 355–369. Masson et Cie, Paris.

Smith, D. S. (1960). Innervation of the fibrillar flight muscle of an insect: *Tenebrio molitor* (Coleoptera). *J. biophys. biochem. Cytol.* **3**, 447–466.

Smith, D. S. (1963). The organization and innervation of the luminescent organ in a firefly, *Photuris pennsylvanica* (Coleoptera). *J. Cell. Biol.* **16**, 323–359.

Smith, D. S., and Treherne, J. E. (1963). Functional aspects of the organization of the insect nervous system. *In* "Advances in Insect Physiology" (J. W. L. Beament, J. E. Treherne, and V. B. Wigglesworth, eds.), Vol. 1, pp. 401–484. Academic Press, London and New York.

Trujillo-Cenóz, O. (1959). Study on the fine structure of the central nervous system of *Pholus labruscoe* (Lepidoptera). *Z. Zellforsch.* **49**, 432–446.

Trujillo-Cenóz, O. (1962). Some aspects of the structural organization of arthropod ganglia. *Z. Zellforsch.* **56**, 649–682.

Usherwood, P. N. R. (1961). Spontaneous miniature potentials from insect muscle fibres. *Nature, Lond.* **191**, 814–815.

Analysis with Microelectrodes of the Synaptic Transmission at the Level of the Sixth Abdominal Ganglion of a Cockroach, *Periplaneta americana*

J. J. CALLEC and J. BOISTEL

Laboratoire de Physiologie Animale,
Faculté des Sciences de Rennes, France

Pumphrey and Rawdon-Smith (1937) and later Roeder (1948) have shown, with anatomical and electrophysiological techniques, that at the level of the sixth abdominal ganglion of a cockroach (*Periplaneta americana*) there is a region of synaptic connexion between the endings of the cercal nerve fibres and those which contribute to form the ascending giant fibres. The characteristics of the presynaptic and postsynaptic responses were described in these studies. More recently, these topics have been investigated by Yamasaki and Narahashi (1960) who were also using external electrodes, the recording ones being placed under the anterior part of the sixth abdominal ganglion. These authors have shown that with this special condition several electrical phenomena occur following stimulation of the cercal nerve, phenomena which were interpreted as occurring in the following order: a cercal potential, a presynaptic potential, and excitatory-postsynaptic potential (e.p.s.p.) and a giant fibres potential. We have resumed this study with microelectrodes with a view to localizing these different phenomena.

I. Techniques

The nerve cord was dissected from the metathoracic ganglion to the sixth abdominal ganglion, including the cercal nerves and their branches inside the cercus. The preparation was put horizontally into a perspex chamber (Fig. 1) which was divided into two parts by a vertical watertight partition, so that the sixth abdominal ganglion and the cercal nerves were located in one compartment covered with physiological solution (Roeder as modified by Boistel, 1957). This solution was constantly oxygenated, which was necessary to maintain the preparation in a satisfactory condition. The cercal nerves were stimulated between a liquid electrode into which they were sucked up, and an AgCl/Ag electrode immersed in the solution. The sixth abdominal ganglion was put ventral side uppermost, the sheath being removed on this surface. A microelectrode could then be inserted into a definite position in the

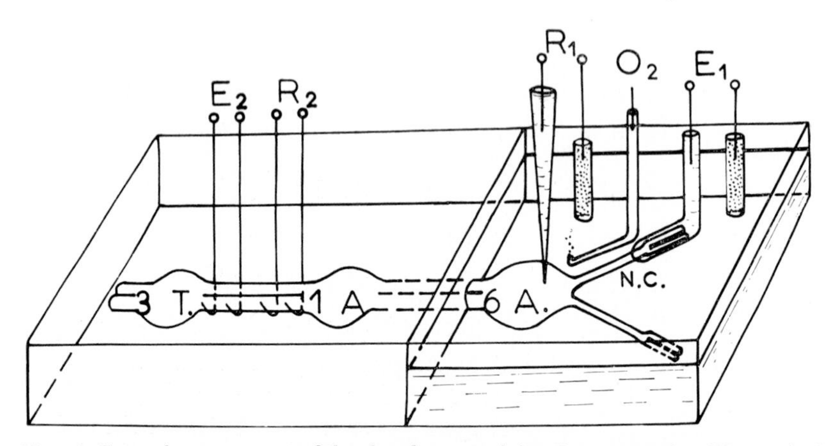

FIG. 1. General arrangement of the chamber containing the preparation. The vertical water-tight partition, situated close to the sixth abdominal ganglion, divides the chamber into two compartments, the right one only being filled with the physiological solution.

R_1: recording electrodes including one microelectrode and one indifferent electrode.
R_2: AgCl/Ag recording electrodes.
E_1: stimulating electrodes including one liquid electrode in which the cercal nerve (N.C.) is sucked up with a syringe and one indifferent electrode.
E_2: AgCl/Ag stimulating electrodes.
O_2: oxygen supply.
3 T: third thoracic ganglion.
1 A, 6 A: first and sixth abdominal ganglia.

ganglion and located with the ocular micrometer of a stereomicroscope, the depth of penetration being measured in μ with a scale incorporated into the micromanipulator supporting the microelectrode. The remaining part of the cord was kept in air in a moist chamber. Two pairs of external electrodes, an excitatory and a recording one, were put between the third thoracic and the first abdominal ganglia.

II. RESULTS

By stimulating one cercal nerve and recording homolaterally at the level of the ganglion, a region has been delimited in which an early potential lasting about 1·2 msec was recorded (Fig. 2 I). This potential is likely to have a cercal origin. Furthermore, in the middle part of this region, we obtained a potential with a particularly small amplitude whose duration could be as long as 6 msec. These characteristics suggest that this potential is electrotonically transmitted from the final ramifications of the cercal fibres and, as suggested by Yamasaki and Narahashi (1960), it seems to correspond to a presynaptic potential.

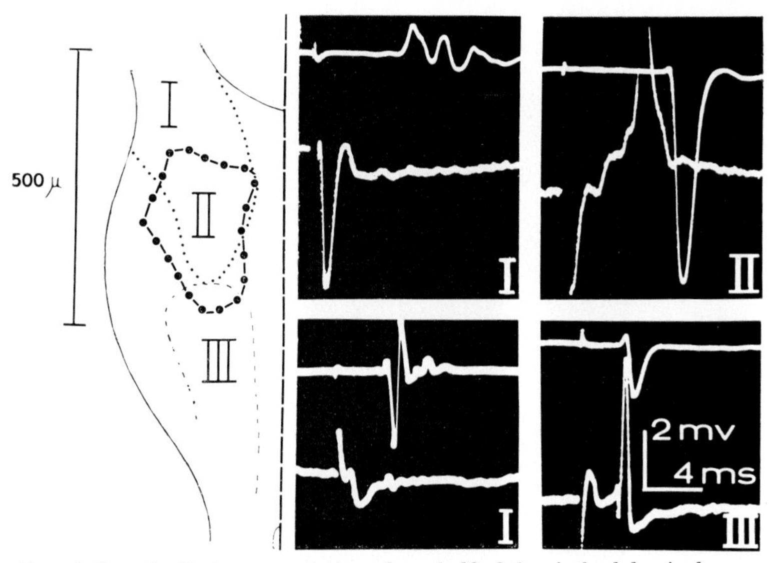

FIG. 2. Longitudinal representation of one-half of the sixth abdominal ganglion, showing in the upper part the insertion of the cercal nerve, and in the lower part the beginning of the connective. Three regions are delimited whose numbers are related to the four graphs on the right side of the figure.

In each division, the lower record was obtained from the corresponding region of the diagram (a downward deflexion being negative); the upper record was obtained from the connectives between the third thoracic and the first abdominal ganglia. These responses were induced by the electrical stimulation of the homolateral cercal nerve.

The vertical scale is related to the lower curve of each photograph, the horizontal one to both curves.

A second region (Fig. 2 II), including the neuropile, described by several authors, Hess (1958), Smith and Treherne (1963), is mainly characterized by potentials occurring after a delay of about 2 msec and lasting approximately 10 msec. For a fixed intensity of stimulation, the ascending phase of this potential generally exhibits a spike which occurs slightly earlier than the propagated action potential recorded at the connective level.

The slow potential, which will be described in detail in the next paragraph, seems to be an e.p.s.p. like those obtained by Yamasaki and Narahashi (1960) with external electrodes. The spike which rises above it is probably transmitted electrotonically from the origin of the giant fibres. Such potentials, but not e.p.s.p. ones, can in fact be recorded inside the whole of zone III (Fig. 2).

If now the potentials obtained inside zone II are investigated in more detail, the following conclusions can be drawn. When the cercal nerve stimulation is progressively increased in intensity, the amplitude of the e.p.s.p. generally increases before the appearance of the propagated

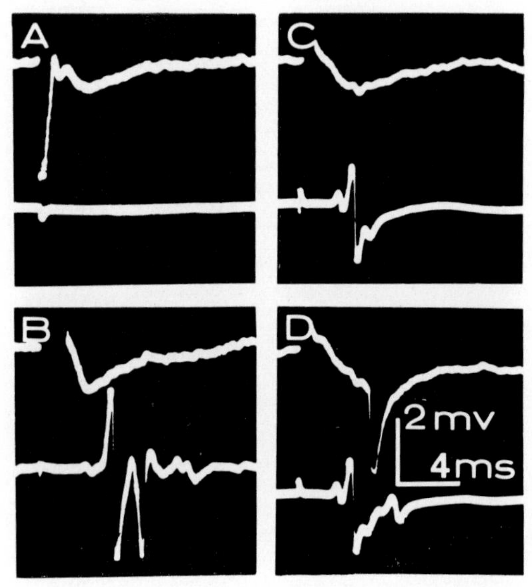

FIG. 3. *Upper records of each graph:* Several types of extracellular responses obtained in region II at a depth of about 50μ at different intensities of stimulation applied to the homolateral cercal nerve.

Lower records: Responses occurring at the connectives between the third thoracic and the first abdominal ganglia.

The vertical scale is related to the upper record of each graph, the horizontal one to both curves.

response. (Figs. 3A, B.) In other cases, it appears simultaneously. This is probably due to the fact that the variations of the stimulation intensity were too extreme. Moreover, in some cases, the e.p.s.p. is topped by the action potential (Fig. 3D) while in others it is not (Fig. 3B). These variations seem to be related to the position of the microelectrode in respect to the synaptic zone itself.

The interpretation of these results is complicated by the fact that they differ according to the depth of the microelectrode tip. The potentials already mentioned generally occur at a depth of about 50μ; at about 300μ (i.e. at approximately two-thirds of the ganglion thickness), they are reversed and their amplitude increases significantly (Fig. 4). In fact, they often release a patterned rhythmical activity (Fig. 5) especially at the slow depolarization phase which precedes the spike itself and which is similar to the one observed when the microelectrode was inserted into a single cell. This record was obtained several times in

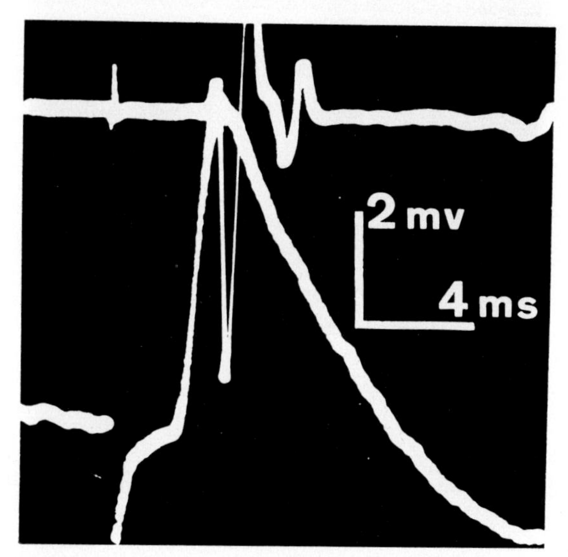

FIG. 4. *Lower record:* "Intracellular" response obtained in region II at a depth of about 300μ, following the stimulation of the corresponding cercal nerve.

Upper record: Corresponding response appearing on the connectives.

The vertical scale concerns the lower curve only.

the very peripheral regions. These potentials are quite similar to those recorded by Tauc (1955) in the ganglion cells of *Aplysia.*

It should be mentioned, however, that the rhythmical activity obtained under these different conditions seems to be caused by the microelectrode penetration, for the frequency of the action potentials progressively decreases and finally this activity stops (Fig. 5a, b). On the other hand, it is generally sufficient to move the microelectrode a little up and down to induce once again the rhythmical activity. As the histological study of the neuropilar region (zone II) does not reveal any cell body, it is likely that the microelectrode penetrated one of the axons forming the giant fibres close to its point of emergence from the

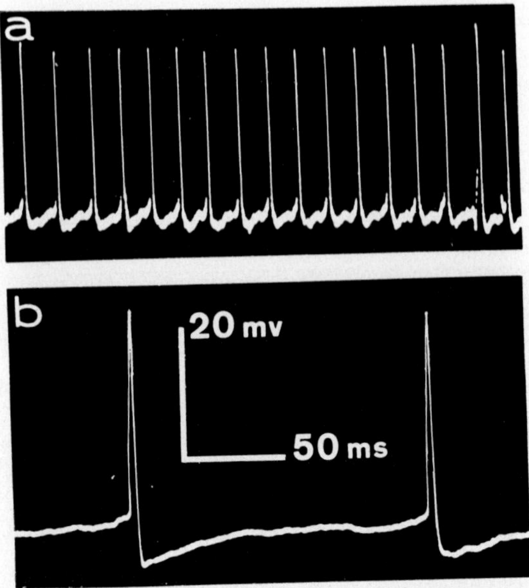

FIG. 5 a. Rhythmical activity recorded at a point very close to the previous one and at the same depth.
b. Same record, about 20 seconds later.

corresponding cell body. The transmission of the electrical phenomenon recorded would thus be an electrotonic one.

III. CONCLUSION

These experiments suggest that it is possible to separate and to localize in the last abdominal ganglion different electrical phenomena involved in synaptic transmission. It is difficult, however, to interpret these preliminary results completely. In the first place we do not know the exact localization of the cell bodies corresponding to the giant fibres although, according to Roeder (1948), they are situated close to the point of entrance of the cercal nerve, probably at the periphery of the ganglion. It is also difficult to localize in the ganglion the origin of the phenomena recorded. In this field, histological studies involving a marking technique and giving accurate information about the position of the tip of the microelectrode are actually in progress in the laboratory.

It would be of some significance to determine whether or not the cell bodies are invaded by the synaptically produced action potential. In a few cases we were able to record, on the connectives, large action potentials which had the same frequency as the ones obtained in single cells with microelectrodes, but which were a little delayed. These preliminary results must be confirmed by antidromic stimulation applied to the ascending giant fibres and recorded in one of the corresponding cell bodies.

It appears that the simultaneous use of electrophysiological and histological techniques will allow a better understanding of the mechanisms involved in synaptic transmission in insects, particularly the influence of temperature and of different pharmacological substances.

REFERENCES

Boistel, J. (1957). "Caractéristiques fonctionnelles des fibres nerveuses et des récepteurs tactiles et olfactifs des insectes." Thèse de Sciences.

Hess, A. (1958). The fine structure of nerve cells and fibers, neuroglia and sheaths of the ganglion chain in the cockroach (*Periplaneta americana*). *J. biophys. biochem. Cytol.* **4,** 731–742.

Pumphrey, R. J., and Rawdon-Smith, A. F. (1937). Synaptic transmission of nervous impulses through the last abdominal ganglion of the cockroach. *Proc. roy. Soc.* (B) **122,** 106–118.

Roeder, K. D. (1948). Organization of the ascending giant fiber system in the cockroach (*Periplaneta americana*). *J. exp. Zool.* **108,** 243–261.

Smith, D. S., and Treherne, J. E. (1963). Functional aspects of the organization of the insect nervous system. *In* "Advances in Insect Physiology" (J. W. L. Beament, J. E. Treherne, and V. B. Wigglesworth, eds.), Vol. 1, pp. 401–484. Academic Press, London and New York.

Tauc, L. (1955). Etude de l'activité élémentaire des cellules du ganglion abdominal de *l'Aplysie*. *J. Physiol. Paris,* **47,** 769–792.

Yamasaki, T., and Narahashi, T. (1960). Synaptic transmission in the last abdominal ganglion of the cockroach. *J. Insect. Physiol.* **4,** 1–13.

The Effects of Temperature Changes Applied to the Cercal Nerves and to the Sixth Abdominal Ganglion of the Cockroach (*Blabera craniifer* Burm.)

J. BERNARD, Y. GAHERY and J. BOISTEL

Laboratoire de Physiologie Animale, Faculté des Sciences de Rennes, France

Several workers have studied variations in the activity of the insect nervous system in connection with temperature changes either by recording modifications in the spontaneous activity of isolated nerve cords (Kerkut and Taylor, 1956 and 1958), by making threshold measurements with external electrodes (Boistel, 1957) or by using microelectrodes in cockroach giant fibres (Bernard, Boistel and Hamon, 1961). Behavioural observations have also yielded some information. The results obtained have, however, often been contradictory. With this in view, we have investigated the variations in the excitability threshold produced by temperature changes, first in the cockroach cercal nerve and then in a preparation which includes the cercal nerves, the sixth abdominal ganglion and the abdominal nerve cord of the same insect. We have been mainly concerned with the speed of temperature variations.

I. Material and Methods

Cockroaches (*Blabera craniifer* Burm.) reared at a constant temperature (24°C) have been used. The cercus, cercal nerves and nerve cord were dissected and introduced into an air-gap chamber similar to the one already used by Boistel (1957). The electrical stimulation was provided by a special stimulator allowing a measurement of the excitability threshold every four seconds. On the second channel of the oscilloscope the corresponding temperatures were simultaneously recorded through a thermistor situated very close to the preparation. Temperature variations, either large and slow (20°C in 90 minutes) or rapid temperature variations, were applied to the whole nerve cord including the cercal nerves, or limited to the sixth abdominal ganglion. In the latter case, the other parts of the preparation were maintained at room temperature.

II. Results

A. *Slow Variations*

At the cercal nerve level, a slow rise in temperature induced an increase in excitability, while a slow decrease in temperature brought about a

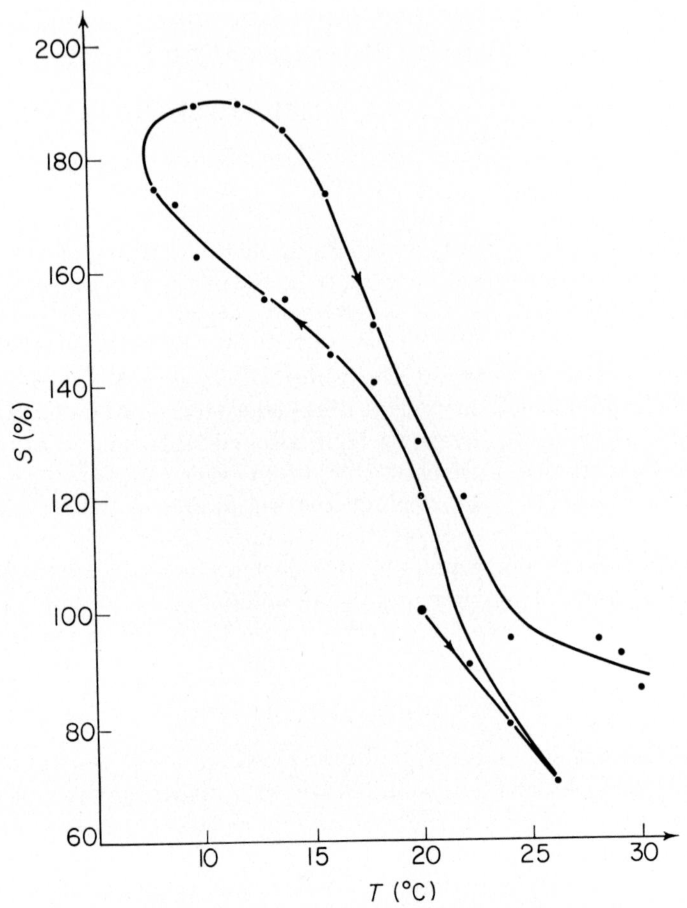

FIG 1. Variations in the excitability threshold of the cercal nerve of *Blabera* following slow temperature changes. The temperature was raised, first from 20°C to 26°C, then it was lowered to 8°C, and lastly raised again to 30°C. The experiment lasted 130 minutes.

decline in excitability (Fig. 1). The threshold value which is 100% at 20°C falls to an average value of 85% at 26°C and to about 90% at 10°C.

In the case of temperature modifications applied to the sixth abdominal ganglion only, records for different temperatures of the ganglion were taken simultaneously on the cercal nerve (upper record) and on the abdominal cord (lower record) after a stimulation of the cercal

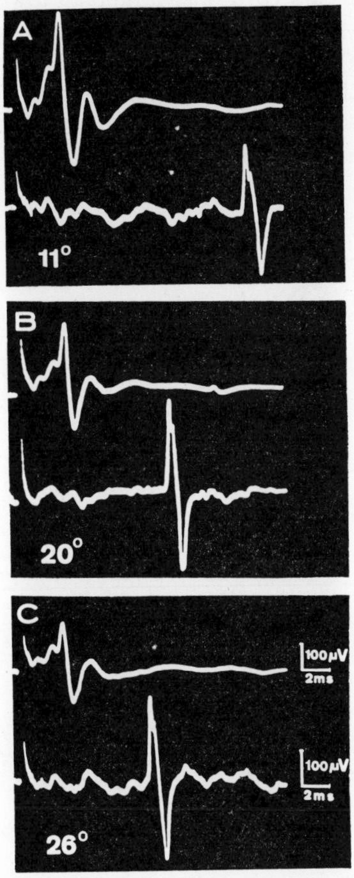

FIG. 2. Modifications in the excitability threshold of the synapses of the sixth abdominal ganglion of the cockroach in function of slow temperature variations. For each photograph:

Upper record: Action potential recorded on the cercal nerve after its electrical stimulation.

Lower record: Action potential of the giant fibres recorded on the cord for the same stimulation.

For the different temperatures each photograph was taken when the first action potential appeared on the cord.

nerve just sufficient to elicit a response on the cord (Fig. 2). Afterwards, on these records, the amplitude of the cercal nerve action potential necessary to induce a synaptic response was measured. It was found that the action potential of the cercal nerve necessary to induce a post-synaptic response on the cord was larger at lower than at high temperatures.

The Q_{10} for these processes had an average value of 1·5 between 15°C and 25°C, so that one can conclude that for slow temperature changes the cercal nerve and the synaptic region are both more excitable at high temperatures than at low ones.

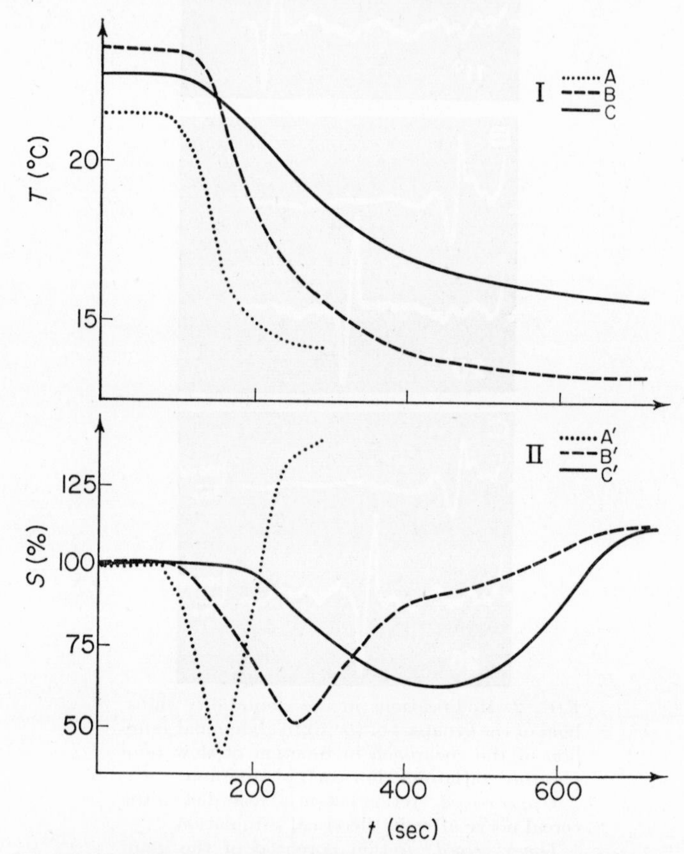

FIG. 3. Variations of the excitability of the cercal nerve of *Blabera*, following different rates of decrease in temperature.

Diagram I (A, B, C) showing variations of temperature in function of time.

Diagram II (A′, B′ ,C′) showing variations of threshold in percent of the threshold obtained at the beginning for the same temperature variations as A, B and C. The three curves were obtained with the same preparation. The temperature is kept at 21°C for about 15 minutes between each of the fast temperature variations.

B. *Rapid Variations*

1. Cercal nerve: The results obtained under these conditions were very different from the previous ones obtained with slow temperature changes. A rapid temperature decrease induced a transient decrease in the thresholds, followed by a gradual increase leading to an inexcitability if the temperature continued to decrease (Fig. 3). The increase of excitability at the beginning of the temperature fall is more important as the variation is faster. On the other hand, a rapidly rising tempera-

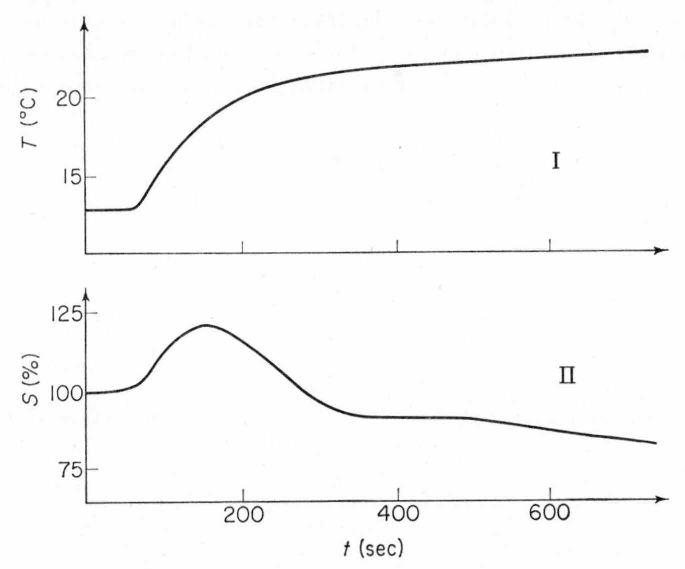

FIG. 4. Variation of the excitability threshold of the cercal nerve of *Blabera* following a temperature increase which is first fast, then slow.

Diagram I showing the relation between the variation of temperature and the time.

Diagram II showing variations of thresholds in percent of the threshold obtained for the initial temperature.

ture provokes a temporary thresholds increase (Fig. 4). In both cases, and as soon as the variation becomes lower, the excitability reverses.

2. Sixth ganglion: A rapid temperature decrease induces a rhythmic activity probably originating in cell bodies or in the synaptic region. If the temperature change is stopped or if the temperature is slowly increased, this activity disappears; a sudden temperature increase again induces the activity. Unfortunately it is impossible, with the present technique, to know exactly which variations in the excitability threshold are occurring.

III. Conclusion

These experiments show that our results differed for slow and rapid temperature variations. They emphasize the importance of the speed of the temperature variations in the functioning of the insect nervous system.

In normal conditions of insect life, the nerve cord, the ganglia and the cercal nerves undergo only slow variations; consequently, the excitability of these tissues must be higher at a higher temperature than at a low one. On the other hand the nerves of the locomotor appendages and especially the different sensilla, thermoreceptors, may undergo such rapid variations in temperature. The influence of these changes on the physiology of these elements will be studied in a subsequent investigation.

References

Bernard, J., Boistel, J., and Hamon, M. (1961). Modifications de l'activité électrique de la fibre nerveuse d'insecte sous l'effet de variations de temperature. *XI Int. Kong. f. Entomologie Wien 1960*, 639–643.

Boistel, J. (1957). "Caractéristiques fonctionnelles des fibres nerveuses et des récepteurs tactiles et olfactifs des insectes." Thèse de Sciences.

Kerkut, G. A., and Taylor, B. J. R. (1956). Effect of temperature on the spontaneous activity from the isolated ganglia of the slug, cockroach and crayfish. *Nature, Lond.* **178**, 426.

Kerkut, G. A., and Taylor, B. J. R. (1958). The effect of temperature changes on the activity of Poikilotherms. *Behaviour* **13**, 259–279.

Study of Some Pharmacological Substances which Modify the Electrical Activity of the Sixth Abdominal Ganglion of the Cockroach, *Periplaneta americana*

Y. GAHERY and J. BOISTEL

Laboratoire de Physiologie Animale,
Faculté des Sciences de Rennes, France

This study was undertaken in order to demonstrate excitatory or inhibitory effects of substances which are able to modify either synaptic transmission or the spontaneous activity originating in the sixth abdominal ganglion.

I. Material and Methods

An isolated preparation has been used, including the cercus, the cercal nerves and the abdominal nerve cord of the cockroach, *Periplaneta americana,* reared at 25°C. Two pairs of electrodes, an excitatory one and a recording one, were applied to one of the cercal nerves. The sixth abdominal ganglion, whose sheath was removed from the dorsal side in order to facilitate the penetration of the substances, was separately and continuously perfused by the physiological solution, with or without the substance to be studied. The effects produced were recorded with external electrodes applied at different places on the nerve cord; fine Ag/AgCl electrodes were used. The whole preparation was introduced into a moist chamber in order to avoid any desiccation and to produce constant moist oxygen supply. This was necessary to maintain the preparation in a satisfactory condition.

II. Results

The following substances were tested: several amino-acids (γ-amino-butyric acid (GABA), β-alanine, α-amino β-oxybutyric acid) and 3-hydroxytyramine.

A. *Amino-acids*

The very important physiological effects of GABA and β-alanine on Vertebrates and Crustacea (Florey, 1960; Boistel and Fatt, 1958; Edwards and Kuffler, 1959; Eisenberg and Hamilton, 1963) have also been studied in a few insect preparations. Milburn and Roeder (1960) did not find any action of these amino-acids on the inhibition of the nerves innervating the phallic musculature of the cockroach. Vereshtchagin, Sytinsky and Tyschenko (1961) pointed out that GABA (10^{-3}) and β-alanine (about 2×10^{-1}), that is to say at high concentrations, had a depressing effect on the electrical activity of the nerve chain of the caterpillar of *Dendrolinus pini*. Suga and Katsuki (1961) have shown that GABA at a concentration of 10^{-2} had an inhibitory effect on the auditory synapses of the locust prothoracic ganglion. Our results obtained on the cockroach confirm this inhibitory effect on the excitatory synapses: this substance at a concentration of 10^{-2} causes a complete inhibition of the response recorded at the level of the ascending giant fibres. A check carried out with intracellular microelectrodes has shown that for this concentration the conduction of the action potential along the giant fibres is not modified by the GABA. Consequently we can conclude that the GABA has a selective action on the synaptic transmission.

Moreover, amino-acids having a chemical structure close to GABA such as β-alanine at a concentration of 5×10^{-3} have a similar effect. However, the α-amino-β-oxybutyric acid has no effect even at a concentration of 10^{-2}.

Further studies will perhaps show how these substances are acting and eventually confirm the interpretations of the inhibitory action of GABA on the excitatory synapses of Vertebrates, given by Curtis and Phillis (1958). According to these authors this substance acts either on the postsynaptic membranes or on the intraneuronic processes which control the membrane properties. It may be possible that the γ-aminobutyric acid plays a role in the normal physiology of the insect. This amino-acid has been obtained by Price (1961) from extracts of the head of a fly, *Musca domestica,* where it appears to be more concentrated than in any other part of the animal. More recently, Ray (1965) has shown that this substance exists in the central nervous system of the cockroach, *Periplaneta americana.* This suggests that in insects also this substance must be involved in the metabolism and physiology of the nervous tissues.

B. 3-*Hydroxytyramine*

Application of this substance to the sixth abdominal ganglion at a concentration of 5×10^{-5} induces after several seconds bursts of activity which are propagated along the nerve cord. This activity ceases a long time after the ganglion is washed with the normal physiological solution (Fig. 1).

The records show that these potentials are of different sizes. However, the ones most frequently obtained can be divided into two categories, the average amplitudes being 0·2 and 1·0 mV respectively. These amplitudes are smaller than those of the potentials observed in

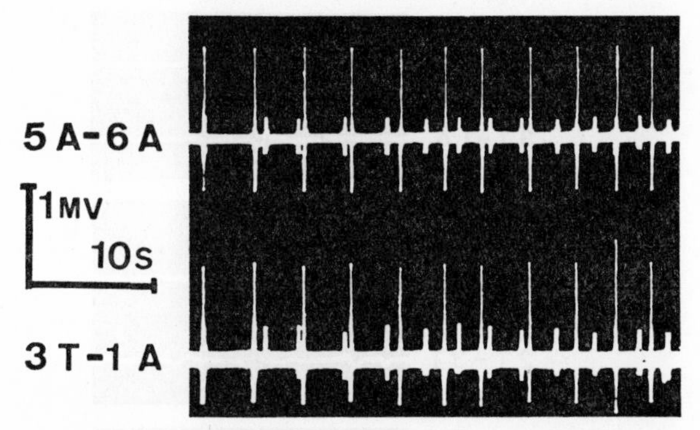

FIG. 1. The 3-hydroxytyramine (5×10^{-5}), applied to the sixth abdominal ganglion, induces bursts of activity which are recorded:
between the fifth and sixth abdominal ganglia (5A–6A);
between the third thoracic and the first abdominal ganglia (3T–1A).

the cord after stimulation of the cercal nerves. The conduction speed is lower than the speed of these action potentials (Fig. 2).

Records carried out at different levels of the nerve cord have shown that it is also possible to record this activity between the third thoracic and the first abdominal ganglion and even between the second and the third thoracic ganglia, although it was impossible to detect any activity beyond the second thoracic ganglion. This seems to show that the fibres involved terminate in this ganglion (Fig. 2).

No discontinuity, concerning either the amplitude or the conduction speed, was observed through the third thoracic ganglion. We can, therefore, conclude that there is no synaptic relay in this ganglion for the fibres already mentioned.

On the other hand, 3-hydroxytyramine does seem to be without effect on synaptic transmission between the cercal nerves and the giant fibres of the cord. No threshold variation could be observed.

According to these different results, the 3-hydroxytyramine induces activity in long fibres, some of which, at least, come to an end in the

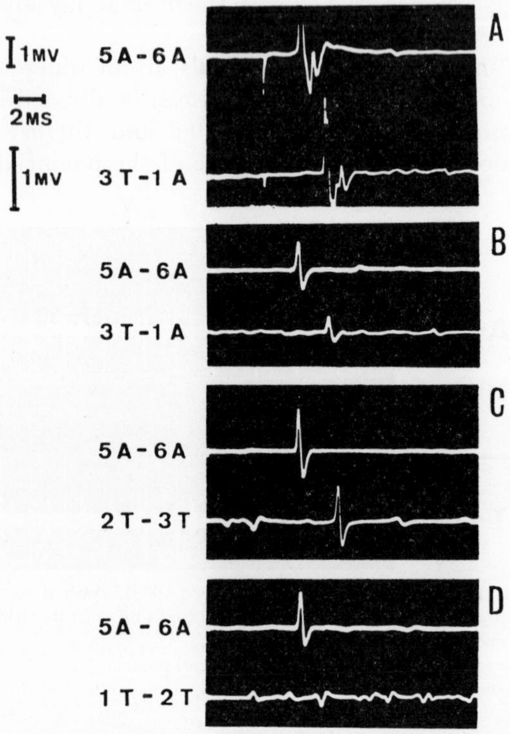

FIG. 2. The action potentials induced by the 3-hydroxytyramine applied to the sixth abdominal ganglion have a lower amplitude and a smaller speed of conduction (B) than those obtained at the level of the giant fibres by the stimulation of the cercal nerves (A). They cross without any discontinuity the third thoracic ganglion (3T; C) but cannot be detected after the second thoracic ganglion (2T; D).

second thoracic ganglion. Moreover, these fibres are different from the ascending giant nerve fibres.

This confirms Hess's work (1958) who, studying the degeneration following the section of the abdominal nerve cord, has shown that the largest fibres (50 to 60μ in diameter) which are related to the cercal nerves reach the third thoracic ganglion, while smaller fibres go further, some eventually reaching the brain.

No fibres have been shown to end in the second thoracic ganglion, although this possibility was not excluded by Hess.

On the other hand, these results are in accord with those obtained by Twarog and Roeder (1957), who recorded bursts of action potentials, which were asynchronous and showing a low amplitude, following the application of adrenaline (epinephrine) and of noradrenaline (norepinephrine) to the sixth abdominal ganglion. This suggests that adrenergic nervous mechanisms may exist in insects; indeed Cameron (1953) has demonstrated by chromatography that an orthodiphenol close to the adrenaline (identical RF) exists in corpora cardiaca extracts of *Periplaneta americana*. Moreover, Östlunde (1954) has found a great amount of 3-hydroxytyramine in insects (from 5 to 10μ g./g.), that is from 10 to 20 times more than the amount of noradrenaline.

REFERENCES

Boistel, J., and Fatt, P. (1958). Membrane permeability change during inhibitory transmitter action in crustacean muscle. *J. Physiol.* **144,** 176–191.

Cameron, M. L. (1953). Secretion of an orthodiphenol in the Corpus cardiacum of the Insect. *Nature, Lond.* **172,** 349–350.

Curtis, D. R., and Phillis, J. W. (1958). Gamma aminobutyric acid and spinal synaptic transmission. *Nature, Lond.* **182,** 323–324.

Edwards, C., and Kuffler, S. W. (1959). The blocking effect of gamma aminobutyric acid and the action of related compounds on single nerve cells. *J. Neurochem.* **4,** 19–30.

Eisenberg, R. S., and Hamilton, D. (1936). Action of γ-aminobutyric acid on *Cancer borealis* muscle. *Nature, Lond.* **198,** 1002–1003.

Florey, E. (1960). Physiological evidence for naturally occurring inhibitory substances. *In* "Inhibition in the Nervous System and γ-Aminobutyric acid". Pergamon Press, Oxford.

Hess, A. (1958). Experimental anatomical studies of pathways in the severed central nerve cord of the cockroach. *J. Morph.* **103,** 479–492.

Milburn, N., and Roeder, K. D. (1960). Chemical control of afferent activity from the sixth abdominal ganglion of the cockroach. *C.R. XI Intern. Kongress für Entom.*, Wien.

Östlunde, E. (1954). The distribution of catecholamines in lower animals and their effects on the heart. *Acta physiol. scand.* **31,** suppl. 112, 1–65.

Price, G. M. (1961). Some aspects of amino-acid metabolism in the adult housefly, *Musca domestica*. *Biochem. J.* **80,** 420–428.

Ray, J. W. (1965). The free amino acid pool of Cockroach (*Periplaneta americana*: Dictyoptera) central nervous system. *In* "The Physiology of the Insect Central Nervous System" (J. E. Treherne and J. W. L. Beament, eds.). Academic Press: London.

Suga, N., and Katsuki, Y. (1961). Pharmacological studies on the auditory synapses in a grasshopper. *J. exp. Biol.* **38,** 759–770.

Twarog, B. M., and Roeder, K. D. (1957). Pharmacological observations on the desheathed last abdominal ganglion of the cockroach. *Ann. ent. Soc. Amer.* **50,** 231–237.

Vereshtchagin, S. M., Sytinsky, I. A., and Tyshchenko, V. P. (1961). The effect of γ-aminobutyric acid and β-alanine on bioelectrical activity of nerve ganglia of the pine moth caterpillar (*Dendrolimus pini*). *J. Insect Physiol.* **6,** 21–25.

Neuronal Pathways in the Insect Central Nervous System

G. M. HUGHES

*Department of Zoology, University of Cambridge**

The study of nervous systems may be approached from widely different points of view, but in all cases questions arise concerning the structure and function of the neurones which form the bulk of this tissue. This is true whether the aim is to investigate the detailed nature of the conduction processes along axons or the neurological basis of the animal's behaviour. Despite a long history of histological work there is still a surprising lack of information about the functional anatomy of insect neurones and there is a danger that advances in this whole field may be held up because of insufficient knowledge about the neuronal architecture. Insects have many advantages as material for studying relationships between structure and function but on the whole their small size is disadvantageous to the neurophysiologist. It is perhaps for this reason that many recent advances in comparative neurophysiology have been made upon larger invertebrates such as crustaceans and molluscs. Nevertheless, these animals are fairly closely related to insects and the application to insects of some of the principles found on them has so far been justified. At the moment the study of the insect central nervous system is in a very rapidly advancing phase because of the utilization of methods and ideas developed on other animals. Consequently many of the views discussed in this paper have received their impetus from knowledge gained with other invertebrates. As the technical difficulties are overcome, some of the advantages of insects as experimental material, such as the possession of a small number of neurones, may result in the discovery of integrative mechanisms that are more widely distributed. It is therefore perhaps a convenient time to summarize some of the techniques that have been used in the investigation of neuronal pathways and to consider the basic connections between neurones so as to indicate some of the gaps in our knowledge of the principles which govern the initiation and transmission of impulses within the insect central nervous system.

* Present address: Department of Zoology, University of Bristol, England.

I. Techniques for the Analysis of Neuronal Pathways

A. *Histological*

The main staining methods for the study of central nervous systems make use of *intra-vitam* methylene blue or impregnation with different metals, especially silver. Methylene blue tends to stain certain cells completely but to leave others unstained. It picks out individual neurones, and the most complete account of any insect central nervous system we owe to Zawarzin (1924a, b) who used this technique. The

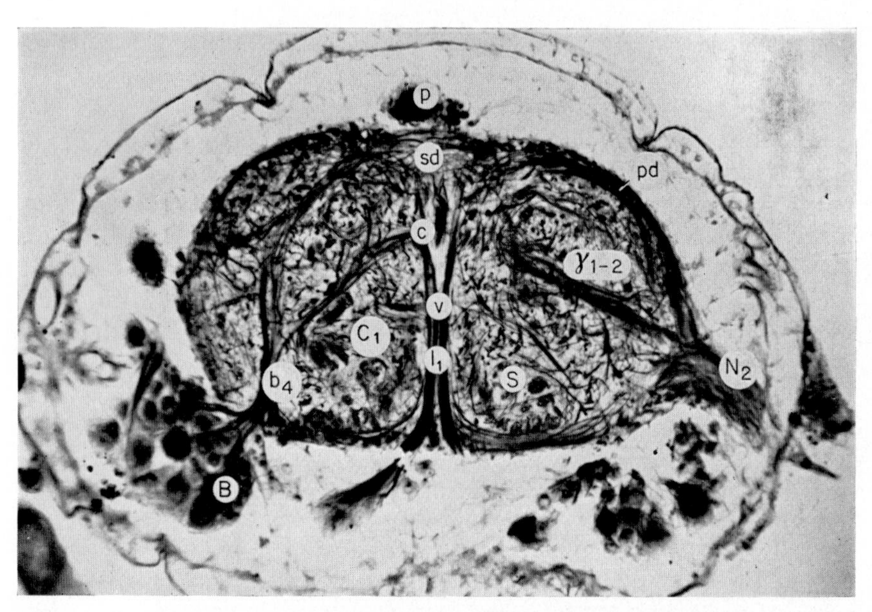

FIG. 1. *Anax* larva. Transverse section through the last abdominal ganglion just behind the origin of the second nerve. The tracts and cell groups are labelled according to the system of Mill (1964): pd, sd, c and v are transverse tracts. Fixation: 0·2% osmic acid in saturated picric acid. Stain: Holmes silver-on-the-slide impregnation method; 15μ sections.

Golgi method of impregnating single neurones with silver has not been applied successfully in many cases although a recent study of Guthrie (1961) is a notable exception. Other silver staining methods give very complete impregnation to all the neurones (Fig. 1) but lead to sections which are very difficult to interpret. The osmium-ethyl gallate method of Wigglesworth (1957) also produces extremely complete impregnation, the results of which are very similar to those obtained with the electron microscope. From the relatively coarse point of view of establishing connections between neurones, electron micrographs give little information although they are very valuable in showing the closeness of the contact between the fine endings within the neuropile. What is required is some technique for producing a three-dimensional reconstruction of the neuropile based upon some of the better silver staining methods, and ultimately those of the electron microscope.

One of the difficulties of the histological method is that it gives little information concerning the function of given neurones although it is usually possible to distinguish motor- and interneurones. But in the finer endings of the neuropile, all axons and their branches appear identical and details about their connections are impossible to obtain at the present time.

B. *Lesion or Ablation Experiments*

This is a more functional method which has been mainly carried out at a relatively crude level, as for example severing whole connectives between individual ganglia. The effects on the animal's ability to respond to sensory stimulation of different parts of the body and the capacity of its muscles to move are observed, and hence information obtained about the pathways present in the cut region. Such work has been done on many insects, a notable example being that of Holst (1934) who found that severing a single connective between two abdominal ganglia of a caterpillar resulted in paralysis of the muscles in the segment posterior to the cut, although this segment remained sensitive, as other parts of the body responded when it was touched. It will be shown later that this paralysis is due to the cutting of a tract of motor axons which descend from the anterior ganglion where they have their synaptic origins. Such a tract has been known for a long time in many different insects as, for example, cockroaches (Binet, 1894), *Corydalis* (Hammar, 1908; Hilton, 1911a, b). It was also shown to be present in the cockroach abdominal nerve cord by electrical stimulation methods when investigating the properties of the abdominal muscles (Hughes and Wilson, V. J., unpublished). Cutting a single connective between the thoracic ganglia has been shown to alter the detailed timing of the fore or hind leg adjacent to the cut (Hughes, 1957). These relatively crude experiments might easily be followed up by finer experiments in which only parts of individual connectives were severed. A further refinement of this technique involves microcautery and has been applied in several instances to the insect brain (Huber, 1964) although not to many of the lower ganglia.

C. *Degeneration*

This technique has been of great value in vertebrate neurology and depends upon the fact that when a nerve is cut the axons contained in it degenerate, particularly when they are no longer connected to the

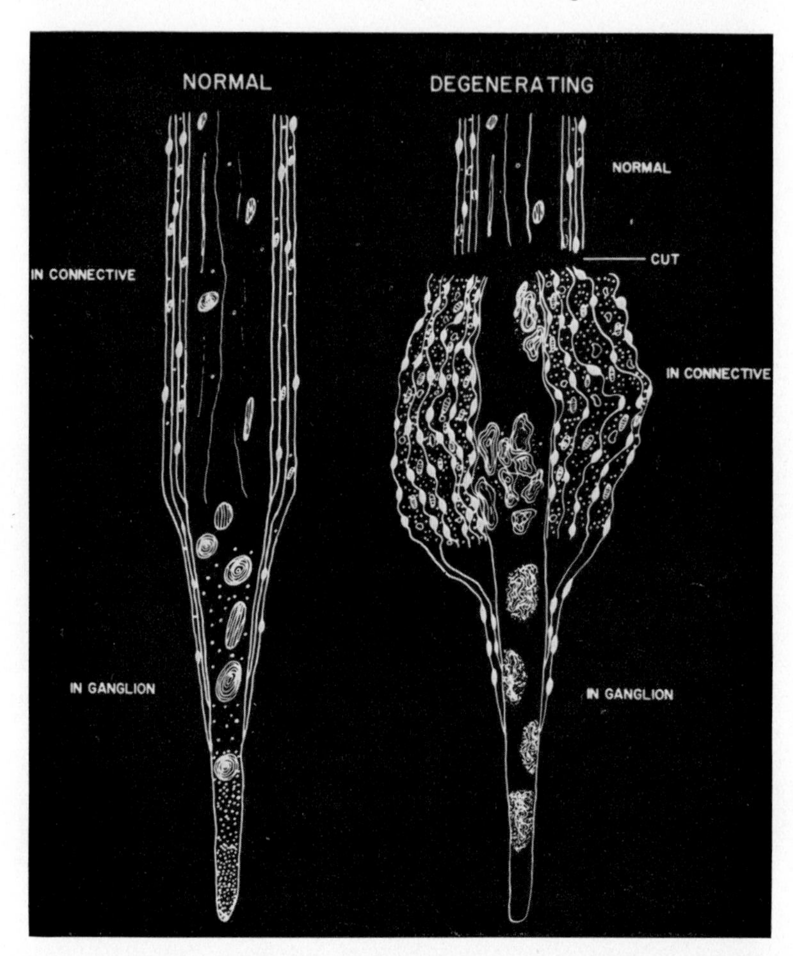

FIG. 2. Diagrams to show the changes during degeneration of a peripheral stump of an insect nerve fibre, its termination and sheath. (From Hess, 1960.)

soma of the neurone. When the cut nerve proximal to the cut is looked at, i.e. towards the soma, there are not such marked histological changes. These changes are illustrated in Fig. 2 which is based on electron microscope studies (Hess, 1960), which showed that particularly the sheaths of a neurone swell up and there are also changes within the

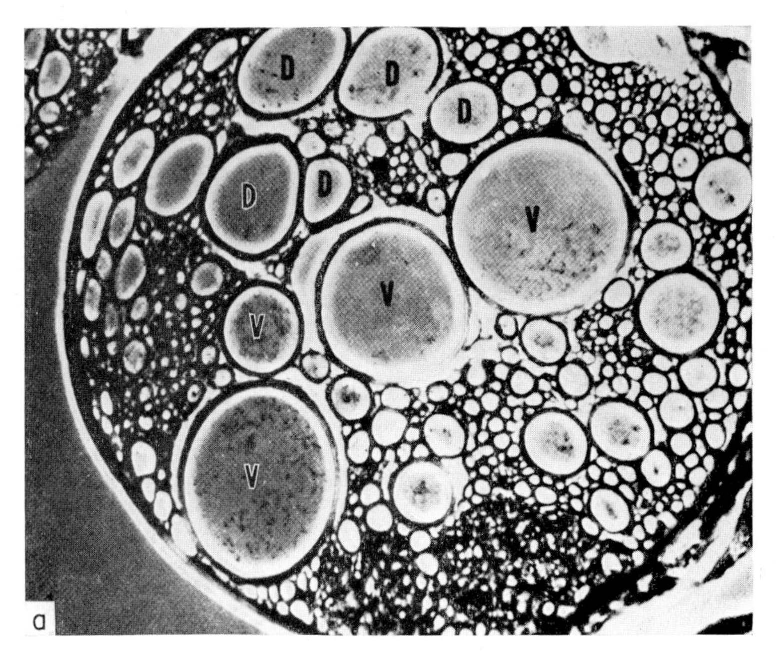

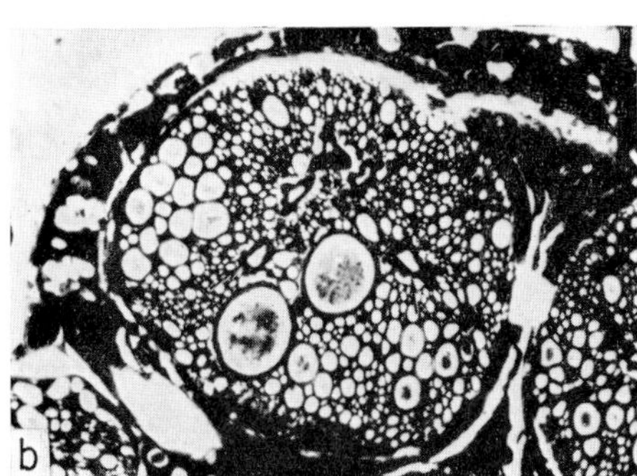

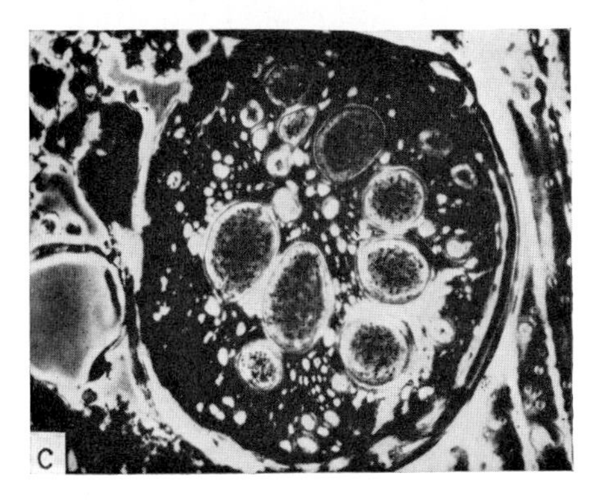

FIG. 3. *Periplaneta*, transverse sections of connectives in the abdominal cord to show changes after it had been cut.

(a) Normal cord showing dorsal (D) and ventral groups (V) of giant fibres.

(b) Anterior to point of severance, four days later.

(c) Posterior to point of severance, six days later. (From Hess, 1958.)

axon. Hess (1958) studied the giant fibre system of the cockroach using this technique. Figure 3 is reproduced from his paper and shows the normal condition of the dorsal and ventral groups of giant fibres. Four days after section of the cord, all except two of the ventral group had degenerated in the connective anterior to the cut. In the connective nearest the last abdominal ganglia we might therefore expect these two giant axons to have degenerated and not the others, but in fact they all remain normal (Fig. 3c). From this and other observations, Hess maintains that the ventral group have their cell bodies in several ganglia of the nerve cord. He considered them to be descending pathways but this is by no means certain because the direction of conduction cannot be judged from the position of the cell body as in vertebrates because the cell body is not the site of synaptic endings. Vowles (1955) had earlier used the degeneration technique to follow individual neurones after destruction of their cell bodies by microcautery. This is made easier in the insect CNS by their cortical position. With this method it is possible to find out the course of the axons of particular cell bodies and in some cases to decide whether they are interneurones or motor neurones.

D. *Electrophysiological Anatomy*

In this method neurophysiological techniques are employed to obtain data of a purely anatomical character but in addition they give information about the types of functional connection between different units of the central nervous system. As will be seen, it emphasizes some features which could not have been appreciated from purely histological investigations. Different methods of recording and stimulation are used, but the information obtained by the use of natural stimulation is of more value than that obtained by electrical stimulation. This is because the natural method excites the pathways by way of the sense organs and so indicates the normal direction of impulse conduction. Furthermore, electrical methods stimulate heterogeneous populations of nerve fibres and have the complication that impulses are made to pass in the opposite (antidromic) direction to the normal.

Various attempts have been made to map different parts of central nervous systems either of the ganglia themselves or of individual nerve trunks. An example of the mapping of a single ganglion is indicated by the microelectrode work of Mill (1963) on the dragonfly larva. Studies of the cross-section of the connective similar to that of Wiersma (1958) might be attempted in the connectives but these are generally too small for studies as complete as those with the crayfish. One of the problems of any mapping method, particularly with microelectrodes, is the identification of the site from which the recordings have been made. The techniques used in vertebrates for marking the position of electrode tip would produce very poor localization in nervous tissues the size of an insect ganglion. The technique for marking individual cells in snail ganglia (Kerkut and Walker, 1962) might be useful for insect central nervous systems.

1. *Recording in mixed nerve trunks*

A basic technique of neurophysiology is the recognition of activity in single axons by recording impulses of a characteristic size, shape and discharge pattern. These may be recognized in recordings from mixed nerve bundles so long as not too many of them are active. In this way it is possible to study a particular region of the body surface following mechanical or other stimulation and so find out whether a given nerve contains sensory fibres from these regions. A survey of this type in the dragonfly nymph did not reveal any striking differences between the

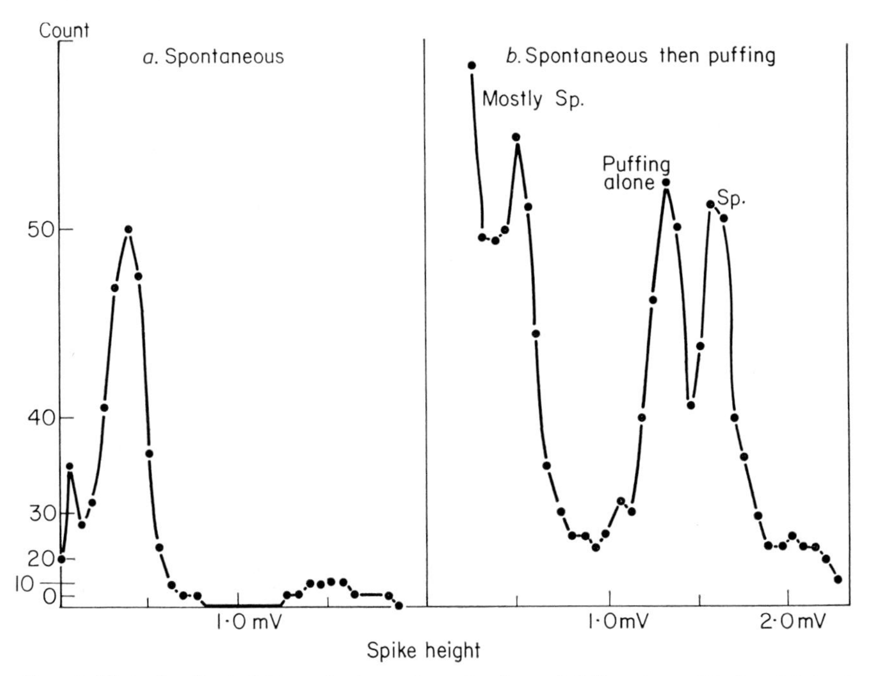

FIG. 4. Plot of spike height against number of spikes of different amplitudes obtained from "kicksorter". The impulses were fed from the nerve cord of a cockroach pinned ventral side uppermost and in (a) shows only spontaneous activity. In (b) air was also blown at the cerci and produced the growth of a peak due to impulses with an amplitude of about 1·3 mV. (Hughes and Ridley, B.W. unpublished.)

external body segmentation and the nervous segmentation of the central nervous system, as had been found in the crayfish (Hughes and Wiersma, 1960).

Relatively little use has been made to date of electronic circuits which classify the impulses in a mixed nerve trunk according to their spike height. The use of devices used by nuclear physicists to plot histograms of spike height ("kicksorters") might well prove of value in further studies of impulse traffic, if they are suitably modified to deal with differences in pulse shape.

Figure 4 shows the results of a preliminary analysis obtained from a cockroach abdominal cord when it was firing spontaneously and following repeated puffs of air to a single cercus. A particularly large group of spikes (1·3 mV) becomes active only during the puffing although another group of large size was active spontaneously. This latter group of large spikes is not always so distinct and they appear to be conducted in a descending direction. Another application of the technique is to obtain histograms showing changes in the pattern of the spike discharges in response to single puffs of air.

2. *Splitting of nerve trunks*

This was first done on peripheral nerves to isolate single axons, but in the study of the central nervous system it has been particularly successful when applied to the connectives between arthropod ganglia. In the abdominal chain of the crayfish it was possible to study the interaction between different integrative regions of individual neurones (Hughes and Wiersma, 1960). The same technique was applied to dragonfly larvae by Fielden and Hughes (1962) and established the presence of sensory fibres ascending and descending in the connectives and was useful in studying the types of interneurones within the central nervous system. This technique alone gives relatively little information about the efferent connections of interneurones, but combined with electrical stimulation of small bundles of axons in the connectives it might provide valuable information. So far it has been used relatively little except in cases such as the crayfish descending command fibres which control the rhythmic movements of the swimmerets (Hughes and Wiersma, 1961).

3. *Electrical stimulation and recording*

In this technique stimulation at one point of the nervous system is followed by recording of action potentials at different places along the course of the pathway and makes it possible to recognize synaptic pathways and also some of the transmission properties along their course. Inherent in this particular method are the disadvantages of electrical stimulation discussed above and the dangers of some of the pathways being non-physiological. A notable example of the application of this method is the analysis of the ascending giant fibre systems of the cockroach (Pumphrey and Rawdon-Smith, 1937; Roeder, 1948) and dragonfly larva (Hughes, 1953; Fielden, 1960).

4. *Microelectrode recording from the central nervous system*

With a sharpened metal electrode insulated except at its tip (Ballintijn, 1961) it is possible to study activity in single units within connectives or ganglia or even a peripheral nerve. Recognition of single units in the dragonfly larva ganglia was obtained by Mill (1963) and a similar technique has also been used by Burtt and Catton (1960) in the optic tract of insects. There are few instances in which intracellular recordings have been obtained, but the work of Hagiwara and Watanabe (1956) on the cicada is a notable exception. They were able to show that a pair of nerve cells in the posterior region of the meso-metathoracic ganglion were the motor neurones of the tymbal muscles. The technique of recording in such cells opens up the possibility of recognizing antidromic potentials following stimulation of a mixed nerve trunk, and so provides evidence for the presence of an axon belonging to the neurone recorded from within the stimulated nerve trunk. The advantages of this technique in tracing pathways in molluscan central nervous system have been illustrated by Hughes and Tauc (1962, 1963). Furthermore, once a microelectrode is within the cell soma it is possible to stimulate the cell directly and, by means of a triggered time base, to recognize impulses in mixed nerve trunks by superimposing many traces. There have been relatively few studies using electrical recording or stimulation of the soma of insect neurones, and many believe that the cell body of arthropod neurones is not excitable. The work of Callec and Boistel (1965) illustrates the use of microelectrodes for the investigation of giant fibre pathways in the last abdominal ganglion of the cockroach, but so far little data has been obtained from recordings within the cell soma.

There are many other techniques that have a subsidiary use in this type of study, as for example the use of nicotine (Pringle, 1939) and other drugs, but they cannot all be mentioned here. The application of the electrophysiological techniques in conjunction with the anatomical ones which constitute the first three categories is of course the ideal approach to the study of pathways. This is being done more and more particularly with the use of microelectrode techniques, and attempts are being made to mark the position of the recording or stimulating electrode within the nervous system. This is extremely laborious work but it is perhaps the best way of improving our knowledge of both the structure and function of these systems at the present moment. It tends to result in what has been described as unit-collecting which has sometimes been compared to stamp-collecting in a derogatory sense. However, we should not be deterred by this description, for like that hobby it can give very important information about the geography, currency, and maybe the rates of exchange between different regions of the central nervous system. This work demands painstaking and sustained study, and those who embark upon it should receive all our encouragement.

II. Different Types of Functional Connections within the Central Nervous System

Within the central nervous system we may distinguish *direct* or through pathways from *synaptic* pathways. The former represent the path of a single axon along which the conduction time is determined by the velocity of impulse propagation. When such pathways are found following electrical stimulation there is the danger that they are not necessarily functional pathways unless the evidence is supported by studies using more physiological inputs. Along synaptic pathways there is at least one region where contact is made between neurones of the same or different types. Direct pathways are characterized by a more constant conduction time and an ability to transmit impulses at higher frequencies. They will conduct impulses in both directions following electrical stimulation, but normally the majority are polarized so that they transmit impulses in one direction only. However, in certain interneurones conduction occurs in both directions under physiological conditions (Fig. 9c). As yet there has not been any clear demonstration of non-polarized synapses within the insect nervous system.

Synaptic connections have been broadly divided into three main types.

1. Integrative synapses in which many sensory impulses are necessary before an action potential is set up in a single post-synaptic element.
2. Non-integrative or relay synapses show a 1 : 1 ratio between input and output.
3. Multiplying synapses are characterized by the production of many postsynaptic impulses in response to a single impulse entering in the synaptic regions.

These terms have sometimes been used a little "loosely" in insect neurophysiology and in the strict sense it has yet to be demonstrated that there is a 1 : 1 synapse between any two single neurones. The same is also true of the demonstration of multiplying synapses. As shown in Fig. 5d, a dragonfly larva interneurone may give a multiple discharge to a single shock applied to a segmental nerve. This, however,

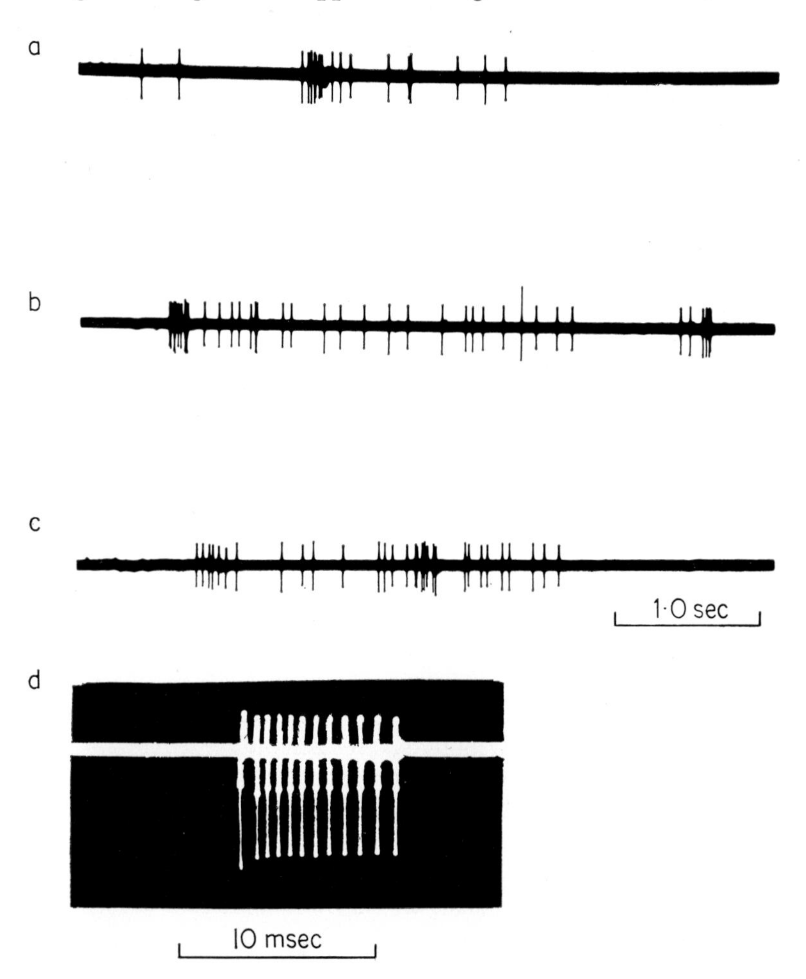

FIG. 5. Electrical recordings from small bundles dissected in the connectives of the dragonfly larva. (a), (b), and (c) are the responses of a multisegmental interneurone in the fourth-fifth connective. The same unit responds to tactile stimulation of contralateral segments (a) 4, (b) 5, and (c) 6 (Fielden and Hughes, unpublished). (d) repetitive firing of a unit dissected in a sixth-seventh connective in response to a single nerve volley to a segmental nerve. (After Fielden, 1963a.)

is the result of the arrival of impulses along many pathways to the neurone that is being recorded from. In no case has an impulse set up in a single presynaptic fibre been shown to lead to multiple responses in a single post-synaptic element. There is, then, a great need for further study of detailed input/output relationships between single neurones of the insect central nervous system.

For convenience, the synaptic pathways within the insect nervous system will be classified according to the different types of neurone involved.

A. *Sensory to Motor Neurone*

This is perhaps the most primitive connection between neurones and it forms the basis of the classical reflex arc. In the vertebrates the afferents and efferents are organized in the dorsal root containing the sensory fibres and in the ventral root in which the motor axons leave the CNS. In the insect it has been suggested that the opposite is true, so that the sensory fibres enter the ganglia in the ventral portions of the mixed nerve, whereas the dorsal parts contain the outgoing fibres. This suggestion was supported by Binet (1894), Hilton (1911) and Zawarzin (1924b) but has frequently been contested by more recent workers, and I certainly thought it sounded too good to be true. Roeder *et al.* (1960) maintain that it does not hold for the cockroach. It has been recently shown to be true at least for the dragonfly larva which was the material upon which Zawarzin based his well-known transverse section. Fielden (1962) split the largest (paraproct) nerve leaving the last abdominal ganglion and showed that the dorsal part contained fibres which conduct the large spikes centripetally and that in this region of the nerve the axons are of large diameter. On the other hand, the ventral portion contains many small axons which conduct impulses of small amplitudes towards the ganglion and are sensory in origin. This demonstration was done entirely electrophysiologically (Fig. 6) and confirms recent morphological work on *Gerris* (Guthrie, 1961), *Rhodnius* (Wigglesworth, 1959) and *Periplaneta* (Pipa *et al.*, 1959).

The role of segmental reflexes in the functioning of the cockroach thoracic ganglia was given an electrophysiological basis by the work of Pringle (1940). More recent studies have shown good preparations for the investigation of segmental reflexes in the dragonfly larva (Fielden, 1963b) and the caterpillar of *Antherea* (Weevers, 1965). As yet there is no information concerning the mechanisms of transmission within the ganglion during these reflexes and it is not even known whether they are monosynaptic. In addition to the evidence for purely segmental reflexes, there is a growing body of knowledge which emphasizes the importance of intersegmental relationships between ganglia. It has been shown, for example, that sensory neurones entering an abdominal ganglion may

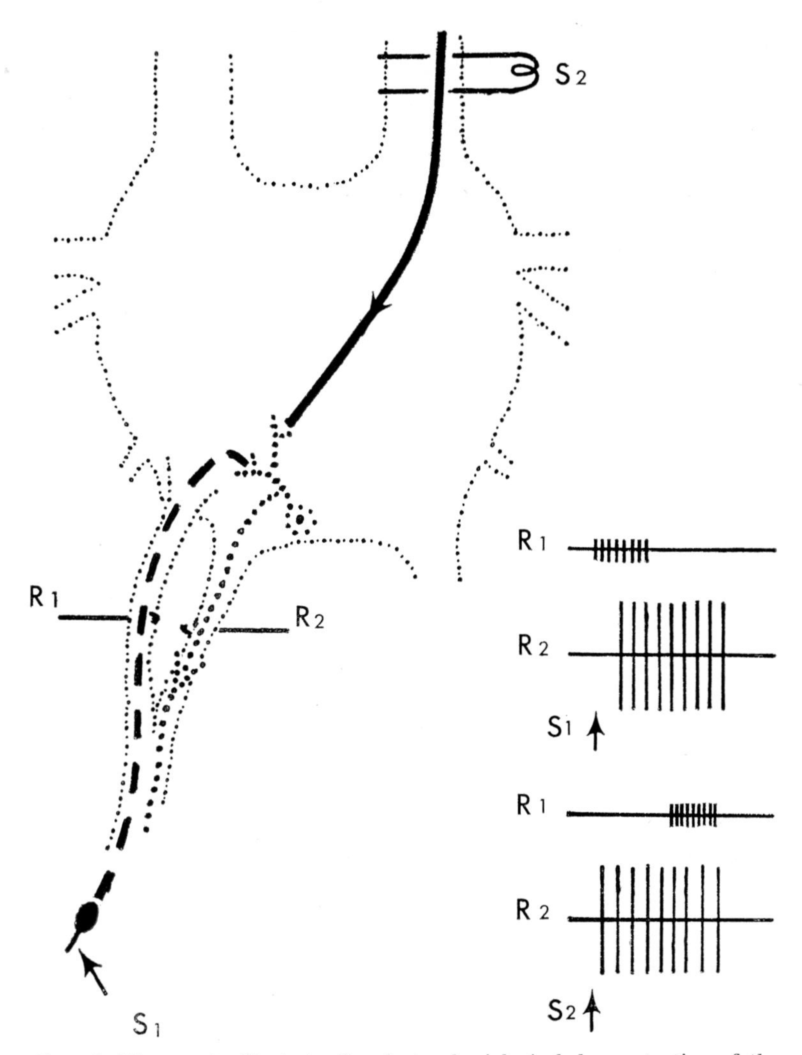

FIG. 6. Diagram to illustrate the electrophysiological demonstration of the separation of sensory and motor axons in the ventral and dorsal parts of a paraproct nerve. The results of mechanical stimulation to the paraproct hairs (S1), and electrical stimulation of the contralateral connective (S2) on recordings from the split nerve indicate the presence of large motor spikes in the dorsal region R1 and small sensory ones in the ventral part R2 (Based on Fielden, 1962b.)

ascend to at least the next anterior ganglion and in some cases such fibres also have connections in the ganglion they enter and in the one posterior (Fig. 7A.1). On the motor side, axons descending from a ganglion in front before emerging in a segmental nerve have been known in

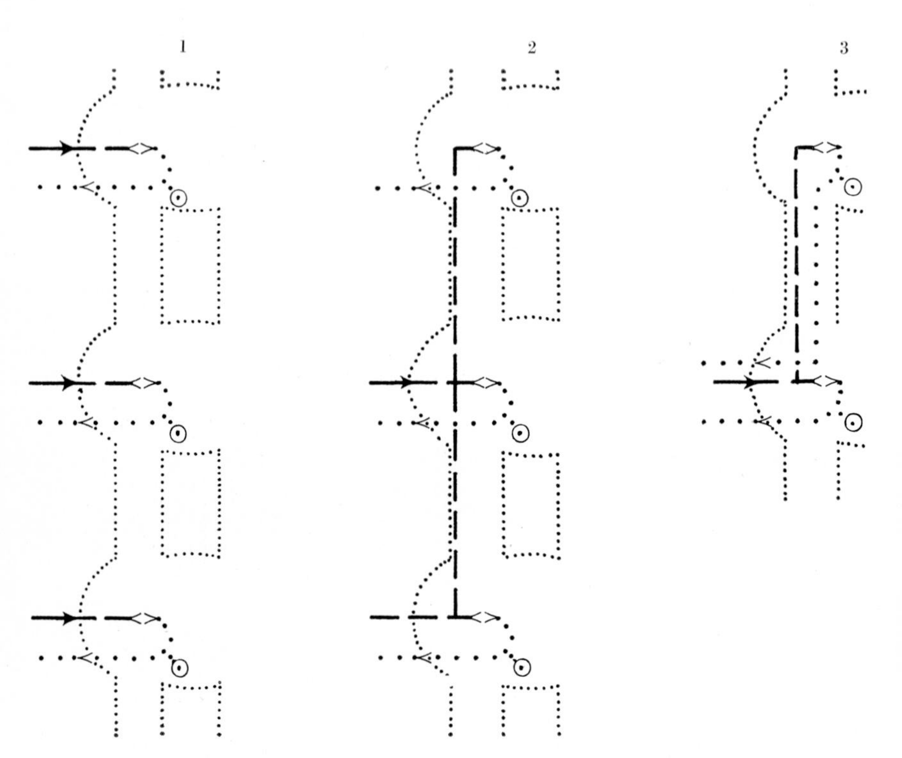

Fig. 7A. Diagrams to illustrate the paths of sensory fibres after entering in a segmental nerve. Dashed lines shows sensory fibres; dotted lines indicate motor fibres.
1 Local endings in segmental ganglion.
2 The sensory axons branch and ascend and descend the cord to synaptic regions in three successive ganglia.
3 The sensory axon ascends the cord and the motor neurone with which it synapses descends before leaving in the same segment as the afferent fibre entered.

several insects and were shown in the cockroach in the abdominal cord of the cockroach for neurones which innervate the rectus abdominis muscle (Hughes and Wilson, V. J., unpublished), and have recently been investigated in detail for the caterpillar of *Antherea* (Weevers, 1965). The nature of one such pathway is represented diagramatically in Fig. 7A.3. An even more complex intersegmental relationship between sensory and motor axons is illustrated by Zawarzin's figure for the motor fibres in the unpaired nerves of the dragonfly larva (Fig. 7B). Here a sensory fibre first ascends to the next anterior ganglion where it

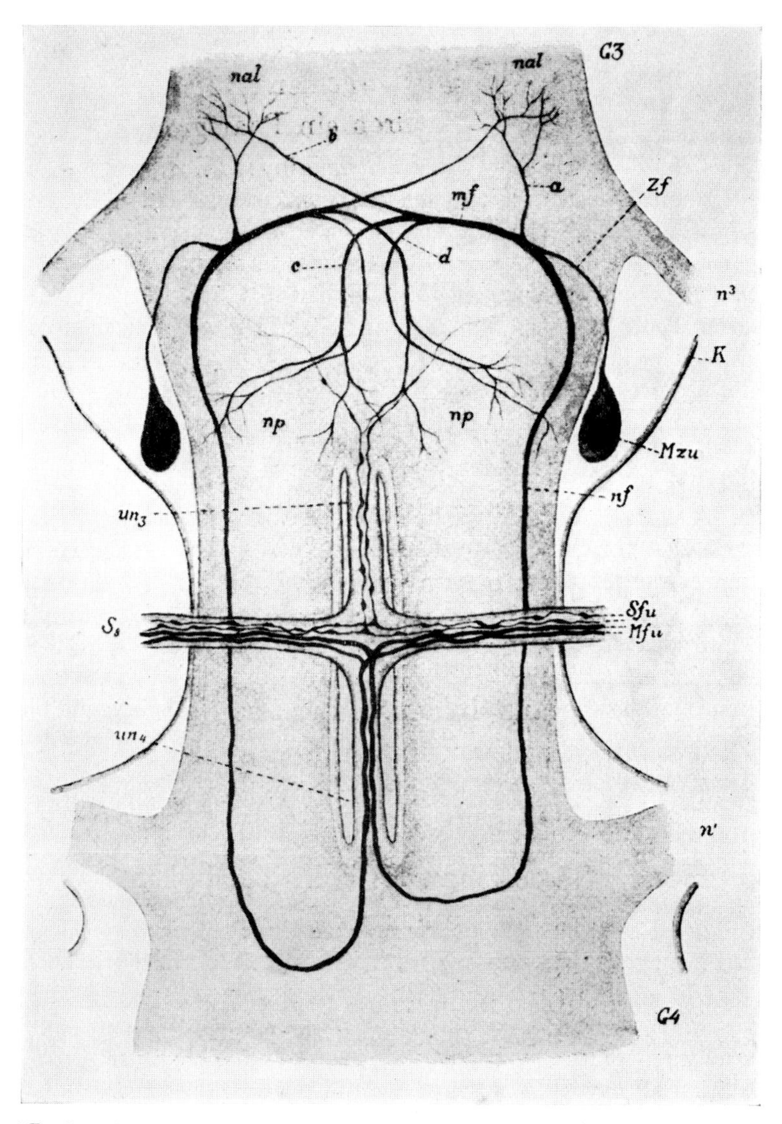

FIG 7B. Complex pathway of sensory and motor fibres in the third and fourth unpaired nerves of a dragonfly larva. (From Zawarzin, 1924a.)

synapses in the neuropile. The motor axon descends in the connective to the ganglion behind and then passes forward to leave in the same unpaired nerve as the sensory fibres entered. This complex relationship has been confirmed by Mill (1964).

B. *Sensory to Interneurone Connections*

Histological methods have shown an abundance of regions where connections can take place between incoming sensory fibres and central neurones, and we are completely baffled by the complexity of the so-called neuropiles in which such connections take place. From such investigations of neuropiles there seems "little possibility of impulses being confined to particular reflex circuits" (Hughes, 1952). Recent work on the neuropile using electron microscopic techniques (Smith and Treherne, 1963) further emphasizes the complexity of the histological picture. There are places where the axons become naked and synaptic vesicles are present in certain regions where two axons approximate to each other. These are presumably the sites of transmitter release, and because of the closeness of their contacts in other regions there is a distinct possibility of electrical influences passing between different axons. The possibility that field effects, either of electrical potential (Hughes, 1952) or chemical transmitters (Roeder, 1958), are involved in the functioning of neuropiles must be borne in mind, but has not yet received any substantial evidence. In other regions of the neuropile the axons are separated from one another by glial elements. As yet it is not known whether completely "insulated" pathways can exist between two neuronal elements, although there are hints of this in some places. As yet there is no evidence that any of the synapses are nonpolarized for vesicles are usually only present on one side of the contact regions. In contrast to this complex picture suggesting considerable interaction between many units, is that obtained as a result of recording from single units dissected in the interganglionic connectives. These units show a surprising specificity in their responsiveness, not only to the modality of peripheral stimulation but also in the distribution of the endings which may excite them. Some interneurones distinguish for instance between different forms of mechanical stimulation applied to spines or to hairs on the body surface or to the abdominal proprioceptors. Such fibres may respond to unisegmental areas but similar units respond to stimulation of several body segments and some to the whole of one side (Fielden and Hughes, 1962). Other fibres respond to regions on both sides of the body and these are the so-called bilateral fibres (Fig. 8b). As in the crayfish there is a tremendous overlap in the central representation of the peripheral input so that stimulation of a given segment of the abdomen, for instance, will excite a very large number of different interneurones. The types of information transmitted by the individual interneurones varies according to their threshold and the nature of their connections. It may be supposed that distant regions of the central nervous system will be affected much more when impulses are transmitted along many pathways simultaneously rather than only coming along a few, and in this way some distinction could be made between relevant information and "noise" within the CNS. In addition to the

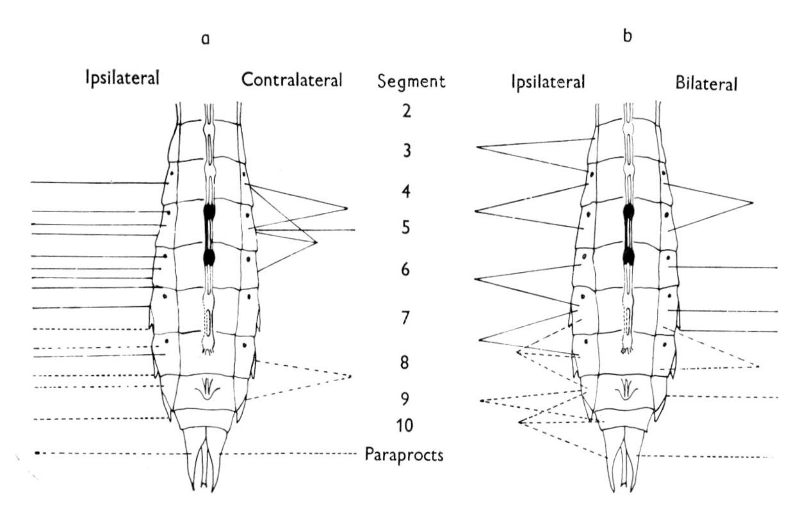

FIG. 8. *Anax* larva. Diagram of ventral abdominal surface to show areas represented by interneurones which have been recorded several times in the right connective between the fourth and fifth ganglia (full lines) or sixth and seventh ganglia (dashed lines). (a) Ipsilateral unisegmental units, and contralateral units. (b) Ipsilateral multisegmental units, and bilateral units. (From Fielden and Hughes, 1962.)

interneurones which respond to stimulation, others were found which appeared to be unaffected by any of the types of stimulation that were used. It is possible that these represent a group of neurones which only respond to more complex patterns of input.

Figure 9 shows four possible patterns of neuronal connections which can interpret the fact that a given interneurone may be excited by stimulation of several body segments. The existence of type c has been demonstrated for many interneurones of the dragonfly larva, and shows that spike initiation may take place in regions spatially separate in segmental ganglia. Because of the presence of sensory axons ascending and descending after their entry to a particular ganglion there is also the possibility of type d (Fig. 9). The demonstration of this type needs more detailed analysis in the dragonfly larva, and its presence has also

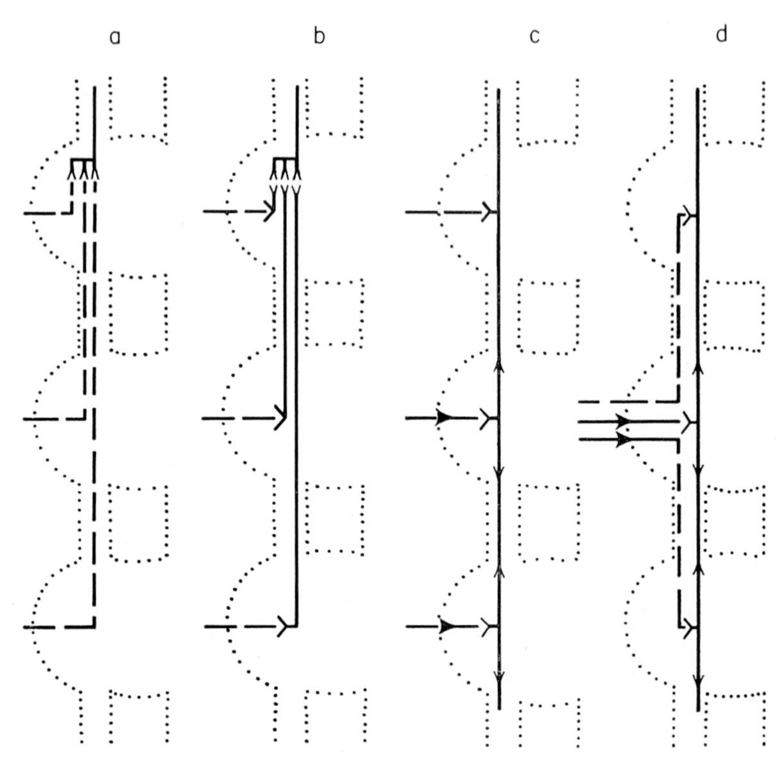

FIG. 9. Diagrams to show the possible types of neuronal connections which would result in a single interneurone firing when sensory areas of three separate segments were stimulated. Dashed lines show primary sensory fibres; full lines interneurones; dotted lines indicate motorneurones. Synapses in which several pre-synaptic fibres converge on to a single postsynaptic fibre are shown—>—, whereas synapses between two single fibres are shown—<>—. The same conventions are adopted in Figs. 4 and 11. The arrows in (c) and (d) indicate the effect of stimulating the sensory inputs entering those particular segments (full arrows); their conduction path along the interneurone is also indicated. It can be seen that simultaneous stimulation of two segments leads to collision of impulses in the interneurone. (Based on Hughes and Wiersma, 1960.)

been indicated in the abdominal cord of the crayfish (Kennedy and Mellon, 1964). Because of interactions due to the collision of impulses (occlusion) the output from either of the types c and d will have a different pattern depending on the direction in which the segments are stimulated successively.

C. *Interneurone to Interneurone Connections*

Relatively little work has been done on such connections partly because of difficulties of isolation as both elements lie within the central nervous system. In the cockroach giant fibre system, the supposed connection between the ascending giants and interneurones which pass forward from the metathoracic ganglion have been shown to transmit in a 1:1 way (Roeder, 1948). However, the more recent histological studies of Pipa *et al.* (1959) throw some doubt on the view that there is in fact such an interruption in the metathoracic ganglion. Many of the bilateral fibres recognized by splitting connectives in the dragonfly larva are excited via another interneurone as has been demonstrated in the crayfish (Wiersma and Hughes, 1961). Further studies in the crayfish have shown the complexity of the second and third order interneuronal connections (Wiersma and Mill, 1965). The whole picture becomes extremely complicated once patterns of interneuronal connections of these types are investigated, and they give some idea of the complexity of the interactions that must occur between different units within the CNS following the arrival of a given sensory input. In the dragonfly larva a connection which is probably between a first and second order interneurone occurs in the next to the last abdominal ganglion and is mentioned by Fielden (1963a). She noted that stimulation of a paraproct nerve excited an interneurone ascending in the ipsilateral connective and that following a brief delay impulses descended the contralateral connective and showed a 1:1 relationship to the ascending discharge. Clearly these two pathways are connected within the ganglion either because the same interneurone sent a branch forward on one side and then backward in the contralateral connective or because there was a 1:1 conducting synapse in the 7th ganglion.

D. *Interneurone to Motor Neurone Connections*

These connections are fundamental in the production of co-ordinated movement patterns but as with the sensory-motor connections they have not been investigated in any real detail. One difficulty is the apparent lability of such synapses, as for example that between the ascending giant fibres and motor neurones in the cockroach metathoracic ganglion. In the dragonfly larva the transmission between ascending fibres and metathoracic motor neurones required fairly frequent (5/sec) stimulation to produce the characteristic folding back of the legs. By contrast stimulation of a thoracic connective produces 1:1 contractions of the leg muscles. Similar 1:1 effects are found in the abdominal muscles following stimulation of the abdominal nerve cord. From these relatively gross experiments it seems possible that there may be 1:1 transmitting synapses between the giant fibres and the abdominal motor neurones, but multiplying synapses between these fibres and the metathoracic motor neurones. In both cases, however, it is possible that convergent synapses are the type involved because many impulses are reaching the motor neurone terminals along several pathways and so long as they arrive more or less synchronously they may lead to the discharge of several impulses in the motor neurones.

Having considered in succession the types of neuronal connection within insect nervous systems an account will now be given of some giant fibre systems as they include connections of several types. Despite a great deal of work there are still many features of these systems that need further study. Particular emphasis will be laid on the analysis of the habituation of the cockroach evasion response as an example of the way in which behaviour of the whole animal may be interpreted at a neurophysiological level.

III. Insect Giant Fibre Systems

Such a system was first discovered in the cockroach nerve cord (Pumphrey and Rawdon-Smith, 1937) and subsequently shown to be present in the locust (Cook, 1951), dragonfly larva (Hughes, 1953), and *Drosophila* (Power, 1948). Although these fibres vary in their diameter they are all large relative to the remaining population of axons. When allowance is made for the size of the animals, the diameters of insect giant fibres are of the same order of magnitude as the giant axons of cephalopod molluscs. The function of these systems in the life of the insect is best known for the cockroach and dragonfly larva and is represented in the flow diagrams of Fig. 10. When a puff of air is directed at the

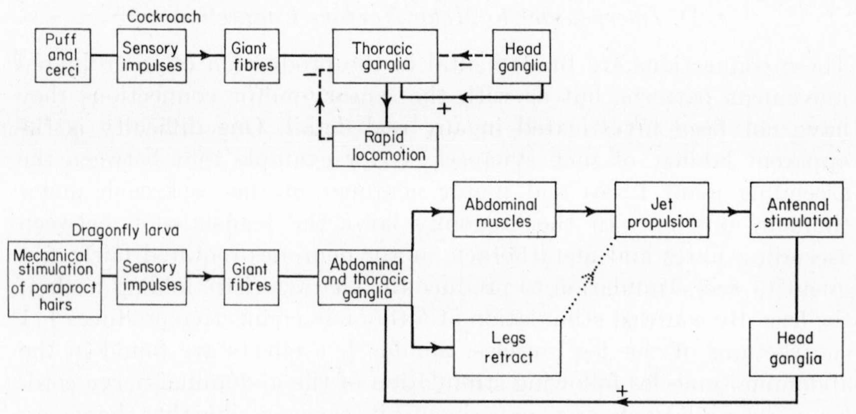

Fig. 10. Flow diagrams to show the effect of eliciting the evasion response in the cockroach and dragonfly larva. The feedback effects from the head are indicated by + and − signs.

anal cerci of a cockroach, sensory impulses are set up in the cercal nerve following stimulation of the fine hairs. These excite the giant fibres having their synaptic origins in the last abdominal ganglion. Large spikes pass forward in the nerve cord to the thoracic ganglia where the motor neurones are excited and produce an increase in the speed of locomotion. The initial response of the legs is probably one of retraction although this may depend upon the posture of the leg concerned. From analyses of cine films it appears that all the legs are excited equally. Once excited this leads to the persistence of rhythmic movements because of the feedback from the normal reflex mechanisms. The thoracic ganglia are also subject to influences from the head ganglia which affect this response. Thus, a decapitated cockroach tends to be hyperexcitable, presumably because of the interruption of some descending inhibitory pathway. However, when the cerci of such an insect are stimulated by a puff of air, the cockroach only makes a short jump and the forward movement is not sustained as it is in the intact insect. This, therefore, indicates some possibly non-specific descending pathway which provides a sort of positive feedback to the thoracic motor neurones. In addition to these parts of the system, there is evidence from various workers that there are some descending pathways in which large impulses are conducted and which may directly fire the giant fibre response. Evidence for such pathways has already been presented (page 86) and has also been noted by Satija (1958a) in *Locusta*, and by Maynard (1956) in the cockroach following stimulation of the antennal nerves. There is a great need for further investigation of the relationship between these descending pathways and those that ascend from the last abdominal ganglion. Certainly it is clear from the more recent histological studies that several of the giant axons pass uninterrupted through the thoracic ganglia where they may well make lateral synaptic connections. In Fig. 11a the various types of neuronal pathways are indicated; the ending of the ascending pathway is shown in the metathoracic ganglion as described by Roeder (1948), but there are certainly some giant neurones passing directly to the other thoracic ganglia.

The corresponding flow diagram for the dragonfly larva is shown in Fig. 11b. The adequate stimulus eliciting jet propulsion is mechanical stimulation of the fine hairs on the inside of the paraprocts. Sensory impulses enter mainly in the largest nerves to the last abdominal ganglion. Here some large axons of the abdominal cord are excited and impulses pass through the abdominal and thoracic ganglia. Unlike the cockroach ascending giant system the impulses make synaptic connection with motor neurones in both the abdominal and thoracic ganglia.

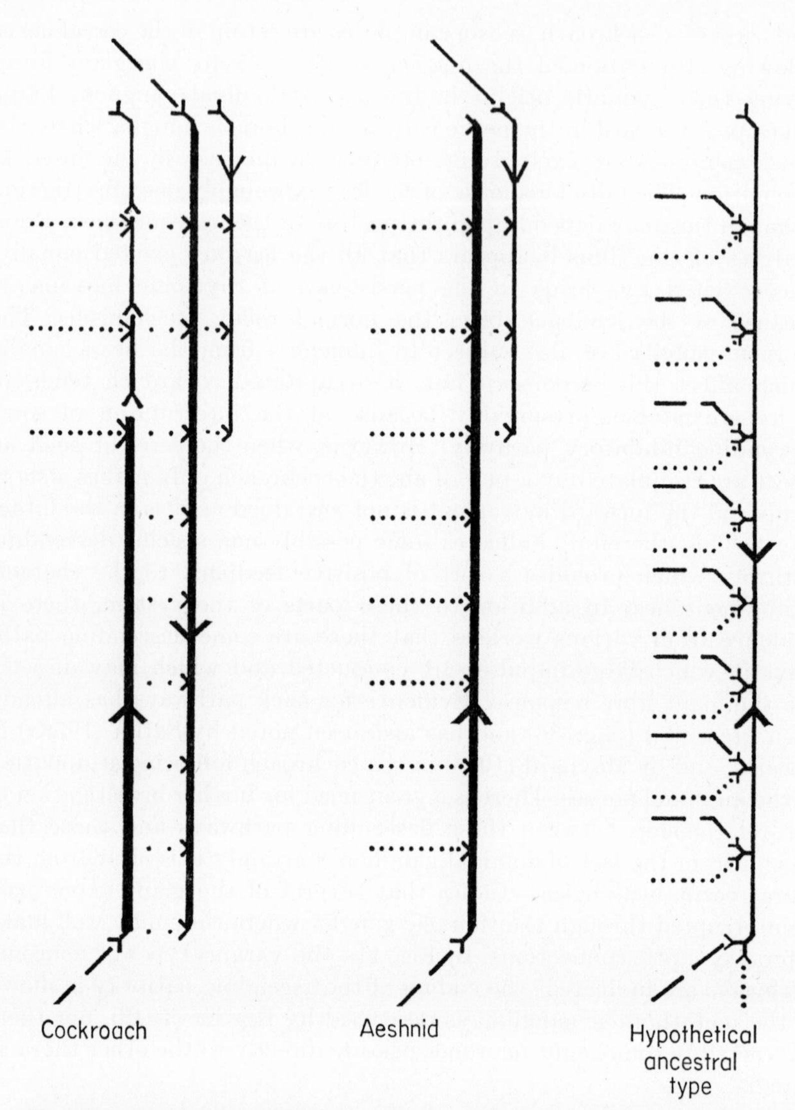

FIG. 11. Diagrams to illustrate some of the neuronal pathways which are involved in the giant fibre systems of (a) cockroach and (b) dragonfly larva. In the cockroach there is some doubt about the presence of the 2nd order interneurones between the metathoracic and the mesothoracic ganglia. Connections of descending giants in the abdominal ganglia have not been established but are well known in the thoracic ganglia. Some non-giant pathways are indicated in both insects and are presumed to conduct the positive or negative feedback effects discussed in the text. In (c), a hypothetical ancestral type is suggested in which a single interneurone has connections in each ganglion with both sensory (dashed) and motor (dotted) neurones. From this type the different sorts of central giant and other interneurones could have differentiated. In both the cockroach and dragonfly larva, the main path of excitation of the giants is from the posterior end. In the dragonfly larva the efferent pathways persist in the abdomen and thorax, but only in the thoracic segments of the cockroach.

Stimulation of the large giants leads to contraction of the abdominal muscles in a 1:1 relationship and produces the jet propulsion of the insect. In the thoracic ganglia the connections produce retraction of all the legs to give a streamlining effect, and may also have a slight propulsive action. Stimulation of the abdominal cord must be at frequencies above 5/sec if retraction of the legs is to take place. Stimulation of a thoracic connective produces a 1:1 retraction with each shock. When the insect is not allowed to move forward there is a retraction stroke of all the legs corresponding to each contraction of the abdominal musculature. Normally, however, the antennae are stimulated during forward progression and this excites some descending pathways which maintain the legs in their retracted position. This might be considered analogous to the descending positive feedback system mentioned in the cockroach.

Electrophysiological investigation of the cockroach system was first carried out by Pumphrey and Rawdon-Smith (1937) who showed that it was necessary to have a certain amount of spatial summation before the post-synaptic response would take place. This was particularly true following previous stimulation, but a fresh, "unadapted" preparation was much more responsive and spatial summation could not be demonstrated. One interesting feature to which Pumphrey and Rawdon-Smith drew attention was what they described as "adaptation" of the transmission across the ganglion. Following repeated submaximal synaptic stimulation of a cercal nerve, they found a decline in the post-synaptic response. This "adapted" state of the ganglion could be broken by an increase in the intensity or frequency of stimulation or by the interpolation of an extra shock in the regular train of stimulation. Furthermore, they frequently obtained a pronounced after-discharge from a preparation when an electrical stimulus was given following a period of rest.

The existence of these phenomena is of interest in relation to the observed decline in the evasion response of a cockroach subject to repeated puffs of air. This first became apparent to me when trying to take some cine films of this response in order to determine whether it consisted of a synchronous extension of all the legs or was a generalized increase in locomotory activity. Analysis of the photographs showed that the latter appears to be the case. Roeder (1959) has analysed the neural pathway of this response and has determined the startle time, i.e. the time from presentation of a puff to the increased electrical response in the metathoracic leg. The time varied from 28–90 msec and he observed a decline in the response with repeated puffing. This he regarded as being due to changes in the transmission within the metathoracic ganglion. Certainly the synapse between the giant fibres and the motor neurones is a particularly labile one as can be demonstrated by recording the action potentials in the fifth nerve or extensor trochanteris muscle following electrical stimulation of the ventral nerve cord or blowing on the anal cerci.

However, in the course of studies of this type it soon became apparent that there are also changes in the giant fibre discharge ascending the abdominal cord following repetition of a constant puff of air. There was no detectable change in the response from the cercal nerve even with quite frequent stimulation (1/sec). The cercal nerve response also remains as a constant asynchronous discharge when the cerci are subjected to a continuous jet of air. Under these conditions the giant fibre discharge in the ventral nerve cord falls off quite rapidly during about a quarter of a minute. In the light of these observations a quantitative study was made of the waning of the response to a constant puff of air as it might be considered a case of habituation in a single ganglion. The preparations used in these experiments were usually reduced to the simplest situation by isolating the abdomen completely and pinning it

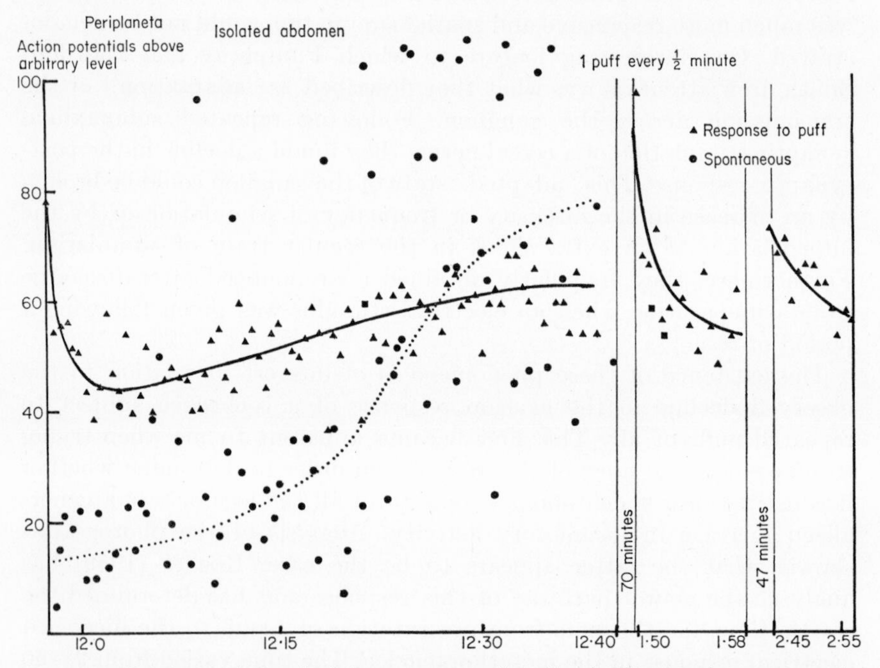

FIG. 12. *Periplaneta*, isolated abdomen preparation (5. xi. 54), showing response to standard puff of air recorded in the ventral nerve cord. The count of spikes above a constant voltage is plotted (▲) for successive puffs given at half-minute intervals. The spontaneous activity of the preparation was also counted between the puffs (●) and the mean level is indicated by the dotted line.

on a waxed block in a moist chamber. Recordings were made from silver/silver chloride electrodes placed under the ventral nerve cord. The impulses discharging to each puff of air were counted using the level selector and scaling unit that had been used in experiments on slug ganglia (Hughes and Kerkut, 1956). The puffs were directed at the cerci by means of a glass tube and remained constant because each time a volume of air filling a given length of rubber tubing to a constant pressure was delivered via an air relay.

When a puff of air is repeated at half-minute intervals there is a decline in the response as shown in Fig. 12. It can also be seen that following a period of rest the response returns to about the same level but falls off again upon further repetition, sometimes more rapidly than in the initial presentation. In this experiment there was a further rest period after which the response was heightened and again fell off.

In other experiments the rest period was maintained constant following a small number of puffs, e.g. groups of 5 puffs, each separated by half-a-minute intervals, were given at half-hourly intervals for more than 12 hours. Figure 13A summarizes the results by giving the impulse count for the first puff of a group together with the average of all five within a group. The effect of the first puff is always greater than the mean value for the five puffs. If one looks at the results of each individual presentation within a group of five puffs, it is clear that the count for the first puff is usually the highest but not invariably so. Figure 13B shows the waning pattern for each group of five given at half-hourly intervals. By this procedure, the total number of puffs given to a preparation has been reduced. It is apparent, however, that there is a steady decline in the total activity of this preparation as well as a decline within each group. Whether this long-term effect is a real phenomenon or represents some fatigue is not clear, but it was found in some other preparations.

The question now arises as to what frequency of puffing can be given without such preparations showing a fall in the count for each puff. At a frequency of one per minute there is some habituation (Fig. 14A), but at a frequency of one every five minutes (Fig. 14B) the response generally remains constant. In some of the preparations counts were also made of the spontaneous activity within the ventral nerve cord between presentations of the puff. In some instances a correlation was found between the response to a puff and the spontaneous activity. Some indication of this is given in Fig. 12 where the gradual rise in the general level of spontaneity accompanies the recovery of the response to a puff of air. However, there was some variability in the results obtained in such experiments, and it is difficult to base any definite

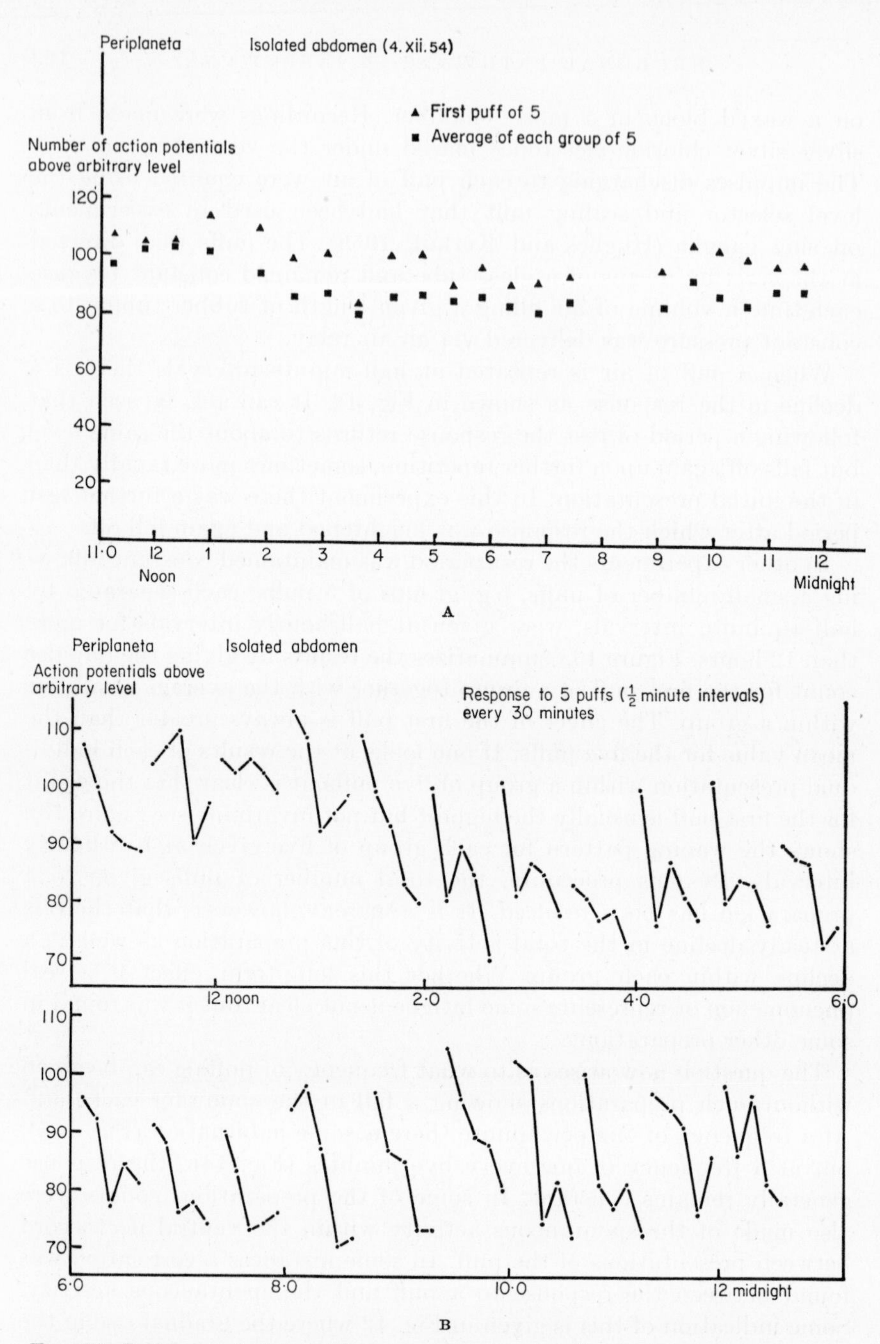

FIG. 13. *Periplaneta*, isolated abdomen preparation (4 xii. 54), showing the counts of impulses recorded in the ventral nerve cord. Repeated groups of five puffs were given every half-hour. The interval between the puffs within each group was half a minute. (A) Plot showing the count for the first puff of a group of five together with the mean figure for each group, (B) details of all the counts within each group of five are shown.

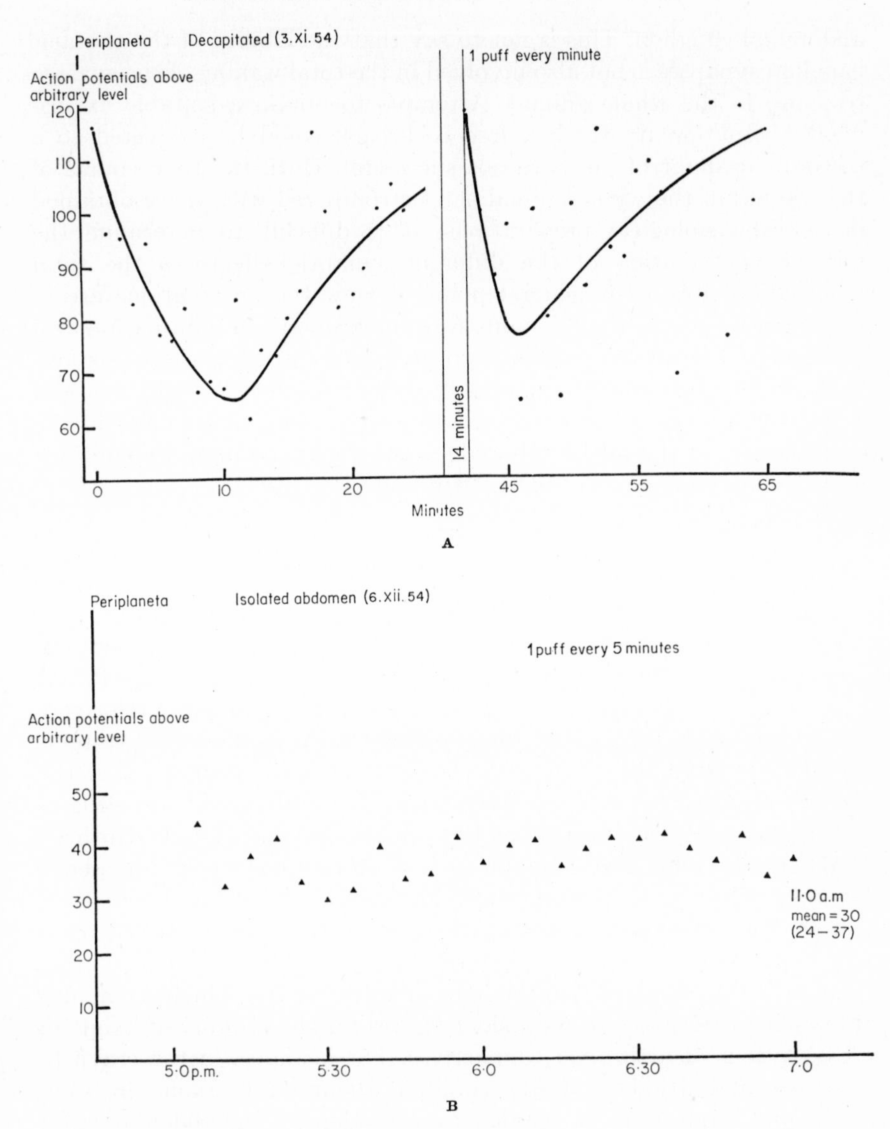

FIG. 14. *Periplaneta*. Decapitated (A) and isolated abdomen preparation (B). Response to standard puff repeated at one-minute and five-minute intervals.

conclusions upon them. In some preparations which showed greater spontaneity than others, the results of repetition of puffs tended not to show such clearcut waning.

It was concluded from these experiments that a part of the mechanism resulting in the waning of the response to a puff of air occurs in the last abdominal ganglion. This is not to say that the lability of the thoracic ganglion synapses is not also involved in the total waning of the evasion response in the whole animal. Attempts to obtain a suitable experimental situation in which a free cockroach could be subjected to a constant draught of air were not successful. Until the time course of the waning in the whole animal can be compared with those obtained in neurophysiological preparations, it is difficult to determine the relative contribution of the different synaptic effects to the total habituation. From the general point of view the observations are of significance because they demonstrate the property of habituation at a neurophysiological level. Such a property of central neurones has also been observed in whole animal preparations of *Aplysia* (Hughes and Tauc, 1962), and has recently been observed from single units in the optic ganglia of the locust (Horridge *et al.* 1964) and in unit recordings from the mammalian mid-brain (Hill and Horn, 1964).

IV. General Discussion of the Number and Properties of Insect Neurones

From the survey presented in this paper it is apparent that there are considerable possibilities for variation in the types of response elicited from the insect central nervous system although it contains relatively few neurones. On the sensory side, however, arthropod nervous systems are characterized by a relatively large number of neurones, e.g. the paired optic ganglia of *Gerris* have about seven times as many cells as the 50,000 found in the rest of the central nervous system (Guthrie, 1961). Each abdominal ganglion of a dragonfly larva contains about 600 cells. Electron micrographs of a locust leg nerve showed that they contain over 2000 axons, mainly sensory, whereas counts based upon light microscopy only gave figures about one-tenth of this number (Rowell, 1964). Similarly, in the light of these results at higher magnifications it is necessary to reconsider figures for the number of axons in cross-sections of arthropod connectives. For instance, in the crayfish, Wiersma and Hughes (1961) counted about 1200 axons in each abdominal connective, but similar cross-sections of the abdominal and circumœsophageal connectives using electronmicroscopy have shown the presence of a very large number of fibres less than 1μ diameter. McAlear *et al.* (1961) have suggested that such counts need to be increased by at least 60%. These numerous fine fibres are probably sensory in function and are doubtless of importance in facilitating the spread of the inputs up and down the cord. It is extremely unlikely that impulses will be recorded from them either using external electrodes on the whole connective or following splitting of the nerve trunks. Dorey and Rowell (1964, personal communication) have also shown in electron-microscope sections that there are very large numbers of small diameter fibres in the locust connectives. As the original counts of fibres in the abdominal connectives of a dragonfly larva (Hughes, 1953) were of the same order as those in the crayfish, this number probably needs to be multiplied by a sizeable factor. Despite these recent studies which indicate the large number of sensory fibres that enter the insect central nervous system, it remains true that the number of central neurones is relatively small. In the dragonfly larva, for example, it is probably only about 600 and in a crayfish about 500/ganglion. For although there are some very small nerve cells it is doubtful whether a significant increase in their number will be established as a result of electron-microscopic studies.

This raises then, the question of how the arthropod central nervous system can co-ordinate such a wide variety of functions with such a relatively small number of neurones. The economy of motor neurones resulting from the oligoaxonic innervation of arthropod muscles is well known, but there must also be important economies at the inter-neuronal level. One of these is clearly the type c interneurone (Fig. 9) which makes lateral synaptic connections within several ganglia. It is of interest to surmise whether there are additional ways which enable a single neurone to fulfil functions frequently ascribed to several as has also been postulated by Bullock (1959). One possibility arises from work done on *Aplysia* neurones (Tauc and Hughes, 1963) where it was shown that transmission between different axonal branches of a neurone may be possible in one way but not in the reverse direction. The failure of an action potential to invade a branch may be due to local differences in membrane properties which affect the excitability, or simply of the relative size of the branches as is used to indicate this property diagrammatically in Fig. 15a. According to this hypothesis spikes may be initiated in the different places as indicated by the arrows. It is evident that different patterns of output from such neurones could be effected in relation to differences of sensory input. For arthropodan inter-neurones the analogous situation might enable portions of a given inter-neurone in a ganglion to function in a similar fashion. Thus in Fig. 15b stimulation of one path (x) gives rise to a localized excitation (1). If another input (y) is excited then the output would be the same as before (1) plus an additional route (2). Only when both paths are stimulated together will the total possible output (1 + 2 + 3 + 4) of the interneurone be excited, including invasion of the main stem of the interneurone which will then be conducted to other segmental levels.

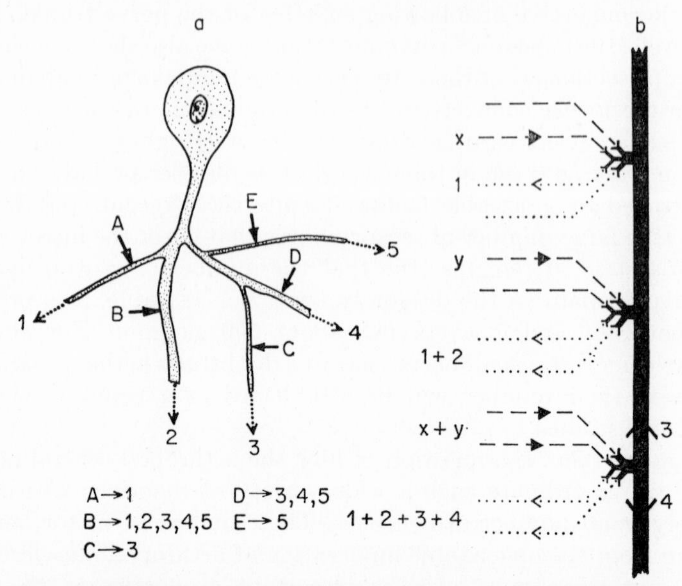

FIG. 15. Diagrams of (a) an *Aplysia* neurone and (b) an arthropod type C interneurone, to show how different parts of individual neurones might function independently of one another. The outputs obtained in response to different inputs are shown in the two cases.

In this way it is possible for a given interneurone to integrate local inputs and to produce variations in output depending upon their nature, as well as to function by integrating between several segments as has been discussed previously. Clearly this is very speculative at the moment, but evidence is accumulating which indicates that the mechanisms discussed for *Aplysia* nerve cells may also apply for neurones in the abdominal ganglia of the crayfish and in the lobster brain.

It seems not unreasonable, therefore, when considering the possible degrees of freedom of an arthropod interneurone to assume in certain cases that conduction between branches can take place one way but not in the opposite direction. There is as yet no evidence for this sort of mechanism from the insect central neurones, but the application of microelectrode recording techniques has only recently begun and information on this question can be expected in the near future. On the other hand it is also known from the supramedullary neurones of the puffer fish that impulses may not be conducted from one branch to another, but that under physiological conditions all branches are invaded (Bennett *et al.*, 1959). On balance, however, it seems justified to explore the possibilities based upon the evidence so far obtained provided it is done with caution until further information is available about the functioning of arthropod neurones under conditions of natural stimulation. From the morphological point of view there is some suggestive evidence. For example the reconstruction by Guthrie (1961) of a giant interneurone of *Gerris* is of extreme interest and is reproduced in Fig. 16. Here the main axon has synaptic connections with several

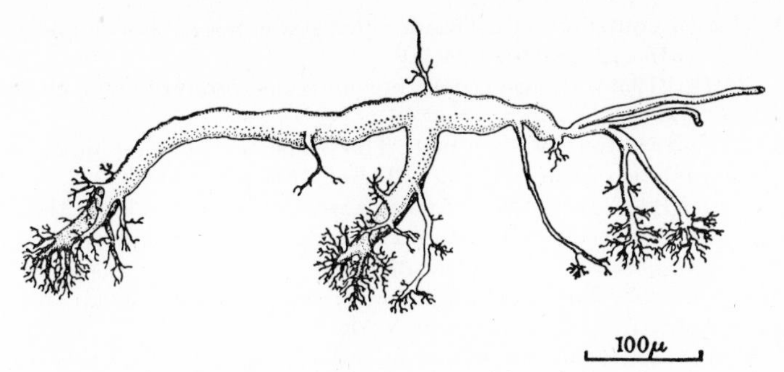

FIG. 16. *Gerris*. The fourth giant internuncial (g, i, 4) as it appears in a preparation of the central nervous system made by the Golgi-Cox technique. The element is viewed from the midline. (From Guthrie, 1961.)

of the ganglionic neuropiles and it is apparent that there are wide variations in the diameter of different branches. It is clearly possible to conceive how local inputs could be integrated by some of these branches without invasion of the main axon and the rest of the regions innervated by the interneurone.

If this concept of interneuronal function turns out to be correct, it will become extremely relevant in any discussion of the number of nerve cells in the insect central nervous system in relation to the integration of complex behaviour patterns. It leads to the view that, at least in certain cases, insect interneurones have greater potentialities for integration than those of mammals, unlike the view expressed by Vowles (1959) which may well be true for other neurones. And, if a single insect neurone can perform functions normally ascribed to a chain of neurones in the mammalian system, it becomes less significant to contrast the performance of the insect brain in relation to that of the mammal solely on the basis of the *number* of nerve cells. As has been emphasized previously this can only be conjecture at the moment, but there is some basis for it from both a morphological and physiological point of view.

REFERENCES

Ballintijn, C. M. (1961). Fine tipped metal microelectrodes with glass insulation. *Experientia* **17**, 523.

Bennett, M. V. L., Crain, S. M., and Grundfest, H. (1959). Electrophysiology of supramedullary neurones in *Sphaeroides maculatus*. *J. gen. Physiol.* **43,** 159–250.

Binet, A. (1894). Contribution à l'étude du système nerveux sous-intestinal des insectes. *J. Anat. Paris* **30**, 449–580.

Bullock, T. H. (1959). Neuron doctrine and electrophysiology. *Science* **129,** 997–1002.

Burtt, E. T., and Catton, W. T. (1960). The properties of single unit discharges in the optic lobe of the locust. *J. Physiol.* **154**, 479–490.

Callec, J. J., and Boistel, J. (1965). Analysis with microelectrodes of the synaptic transmission at the level of the sixth abdominal ganglion of a cockroach (*Periplaneta americana*). *In* "The Physiology of the Insect Central Nervous System" (J. E. Treherne and J. W. L. Beament, eds.), pp. 59–65, Academic Press, London and New York.

Cook, P. M. (1951). Observations on giant fibres of the nervous system of *Locusta migratoria*. *Quart. J. micr. Sci.* **92**, 297–305.

Fielden, A. (1960). Transmission through the last abdominal ganglion of the dragonfly nymph, *Anax imperator*. *J. exp. Biol.* **37**, 832–844.

Fielden, A. (1963a). Properties of interneurones in the abdominal nerve cord of a dragonfly nymph. *J. exp. Biol.* **40,** 541–552.

Fielden, A. (1963b). The localization of function in the root of an insect segmental nerve. *J. exp. Biol.* **40,** 553–561.

Fielden, A., and Hughes, G. M. (1962). Unit activity in the abdominal nerve cord of a dragonfly nymph. *J. exp. Biol.* **39,** 31–44.

Guthrie, D. M. (1961). The anatomy of the nervous system in the genus *Gerris* (Hemiptera-Heteroptera). *Phil. Trans.* B, **244,** 65–102.

Hagiwara, S., and Watanabe, A. (1956). Discharges in motorneurons of cicada. *J. cell. comp. Physiol.* **47,** 415–428.

Hammar, A. G. (1908). On the nervous system of *Corydalis cornuta*. *Ann. ent. Soc. Amer.* **1,** 105–127.

Hess, A. (1958). Experimental anatomical studies of pathways in the severed central nerve cord of the cockroach. *J. Morph.* **103,** 479–502.

Hess, A. (1960). The fine structure of degenerating nerve fibres, their sheaths and their terminations in the central nerve cord of the cockroach (*Periplaneta americana*). *J. biophys. biochem. Cytol.* **7,** 339–344.

Hill, R. M., and Horn, G. (1964). Responsiveness to sensory stimulation of cells in the rabbit mid-brain. *J. Physiol.* **24,** P.

Hilton, W. A. (1911a). The structure of the central nervous system of *Corydalis* larva. *Ann. ent. Soc. Amer.* **4,** 219–256.

Hilton, W. A. (1911b). Some remarks on the motor and sensory tracts of insects. *J. comp. Neurol.* **21,** 383–391.

Holst, E. von (1934). Motorische und tonische Erregung und ihr Bahnenverlaut bei Lepidopterenlarven. *Z. vergl. Physiol.* **21,** 395–414.

Horridge, G. A., Scholes, J. H., Shaw, S., and Tunstall, J. (1965). Extracellular recordings from single neurones in the optic lobe and brain of the locust. *In* "The Physiology of the Insect Central Nervous System" (J. E. Treherne and J. W. L. Beament, eds.), pp. 165–202, Academic Press, London and New York.

Huber, F. (1965). Brain controlled instinctive behaviour in Orthopterans. *In* "The Physiology of the Insect Central Nervous System" (J. E. Treherne and J. W. L. Beament, eds.), pp. 233–246, Academic Press, London and New York.

Hughes, G. M. (1952). Differential effects of direct current on insect ganglia. *J. exp. Biol.* **29,** 387–402.

Hughes, G. M. (1953). "Giant" fibres in dragonfly nymphs. *Nature, Lond.* **171,** 87.

Hughes, G. M. (1957). The coordination of insect movements. II. The effect of limb amputation and the cutting of commissures in the cockroach (*Blatta orientalis*). *J. exp. Biol.* **34,** 306–333.

Hughes, G. M., and Kerkut, G. A. (1956). Electrical activity in a slug ganglion in relation to the concentration of Locke solution. *J. exp. Biol.* **33,** 282–294.

Hughes, G. M., and Tauc, L. (1962). Some aspects of the organisation of central nervous pathways in *Aplysia depilans. J. exp. Biol.* **39,** 45–69.

Hughes, G. M., and Tauc, L. (1963). An electrophysiological study of the anatomical relations of two giant nerve cells in *Aplysia depilans. J. exp. Biol.* **45,** 469–486.

Hughes, G. M., and Wiersma, C. A. G. (1960). Neuronal pathways and synaptic connexions in the abdominal cord of the crayfish. *J. exp. Biol.* **37,** 291–307.

Hughes, G. M., and Wiersma, C. A. G. (1960). The co-ordination of swimmeret movements in the crayfish, *Procambarus Clarkii* (Girard). *J. exp. Biol.* **37,** 657–670.

Kennedy, D., and Mellon, D. (1964). Receptive-field organization and response patterns in neurons with spatially distributed input. *In* "Neural Theory and Modeling", Stanford University Press, California.

Kerkut, G. A., and Walker, R. J. (1962). Marking individual nerve cells through electrophoresis of ferrocyanide from a microelectrode. *Stain Tech.* **37,** 217–219.

McAlear, J., Camougis, G., and Thibodean, L. F. (1961). Mapping of large areas with the electron microscope. *J. biophys. biochem. Cytol.* **10,** 133–135.

Maynard, D. M. (1956). Electrical activity in the cockroach cerebrum. *Nature, Lond.* **177,** 529–530.

Mill, P. J. (1963). Neural activity in the abdominal nervous system of aeschnid nymphs. *Comp. Biochem. Physiol.* **8,** 83–98.

Mill, P. J. (1964). The structure of the abdominal nervous system of aeschnid nymphs. *J. comp. Neurol.* **122,** 157–171.

Pipa, R. L., Cook, E. F., and Richards, A. G. (1959). Studies on the hexapod nervous system. II. The histology of the thoracic ganglia of the adult cockroach, *Periplaneta americana* (L.). *J. comp. Neurol.* **113,** 401–423.

Power, M. E. (1948). The thoracico-abdominal nervous system of an adult insect. *Drosophila melanogaster. J. comp. Neurol.* **88,** 347–409.

Pringle, J. W. S. (1939). The motor mechanism of the insect leg. *J. exp. Biol.* **16,** 220–231.

Pringle, J. W. S. (1940). The reflex mechanism of the insect leg. *J. exp. Biol.* **17,** 8–17.

Pumphrey, R. J., and Rawdon-Smith, A. F. (1937). Synaptic transmission of nerve impulses through the last abdominal ganglion of the cockroach. *Proc. roy. Soc.* B. **122,** 106–118.

Roeder, K. D. (1948). Organization of the ascending giant fiber system in the cockroach (*Periplaneta americana*). *J. exp. Biol.* **108,** 243–262.

Roeder, K. D. (1958). The nervous system. *Annu. Rev. Ent.* **3,** 1–18.

Roeder, K. D. (1959). A physiological approach to the relation between prey and predator. *Smithson. misc. Coll.* **137,** 287–306.

Roeder, K. D., Tozian, L., and Weiant, E. A. (1960). Endogenous nerve activity and behaviour in the mantis and cockroach. *J. Insect Physiol.*, **4,** 45–62.

Rowell, C. H. F. (1964). Central control of an insect segmental reflex. I. Inhibition by different parts of the central nervous system. *J. exp. Biol.* **41,** 559–572.

Satija, R. C. (1958a). A histological and experimental study of nervous pathways in the brain and thoracic nerve cord of *Locusta migratoria migratoriodes* (R. & F.). *Research Bull. Panjab Univ.* **137,** 13–32.

Satija, R. C. (1958b). A histological study of the brain and thoracic nerve cord of *Aeschna* nymphs with special reference to descending nervous pathways. *Research Bull. Panjab Univ.* **138,** 33–47.

Smith, D. S., and Treherne, J. E. (1963). Functional aspects of the organization of the insect nervous system. *In* "Advances in Insect Physiology" (J. E. Treherne, J. W. L. Beament and V. B. Wigglesworth, eds.), Vol. 1, pp. 401–484, Academic Press, London and New York.

Tauc, L., and Hughes, G. M. (1963). Modes of initiation and propagation of spikes in the branching axons of molluscan central neurones. *J. gen. Physiol.* **46,** 533–549.

Vowles, D. W. (1955). The structure and connexions of the corpora pedunculata in bees and ants. *Quart. J. micr. Sci.* **96,** 239–255.

Vowles, D. W. (1961). Neural mechanism in insect behaviour. *In* "Current Problems in Animal Behaviour" (W. H. Thorpe and O. L. Zangwill, eds.), Cambridge University Press.

Weevers, R. de G. (1965). Proprioceptive reflexes and the co-ordination of locomotion in the caterpillar of *Antherea pernyi* (Lepidoptera). *In* "The Physiology of the Insect Central Nervous System" (J. E. Treherne and J. W. L. Beament, eds.), pp. 113–124, Academic Press, London and New York.

Wiersma, C. A. G. (1958). On the functional connections of single units in the central nervous system of the crayfish, *Procambarus clarkii* (Girard). *J. comp. Neurol.* **110,** 421–72.

Wiersma, C. A. G., and Hughes, G. M. (1961). On the functional anatomy of neuronal units in the abdominal cord of the crayfish. *Procambarus clarkii* (Girard). *J. comp. Neurol.* **116,** 209–228.

Wiersma, C. A. G., and Mill, P. J. (1965). "Descending" neuronal units in the commissure of the crayfish central nervous system; and their integration of visual, tactile and proprioceptive stimuli. *J. comp Neurol.* (In press.)

Wigglesworth, V. B. (1957). The use of osmium in the fixation and staining of tissues, *Proc. roy. Soc.* B, **147,** 185–199.

Wigglesworth, V. B. (1959). The histology of the nervous system of an insect, *Rhodnius prolixus* (Hemiptera). II. The central ganglia. *Quart. J. micr. Sci.* **100,** 299–313.

Zawarzin, A. (1924a). Über die histologische Beschaffenheit des unpaaren ventralen Nervs der Insekten (Histologische Studien über Insekten V.). *Z. wiss. Zool.* **122,** 97–115.

Zawarzin, A. (1924b). Zur Morphologie der Nervenzentren. Das Bauchmark der Insekten. Ein Beitrag zur vergleichenden Histologie (Histologische Studien über Insekten VI.). *Z. wiss Zool.* **122,** 323–424.

Proprioceptive Reflexes and the Co-ordination of Locomotion in the Caterpillar of *Antheraea pernyi* (Lepidoptera)

R. DE G. WEEVERS

Department of Zoology, University of Cambridge

I. INTRODUCTION

During the last twelve years, information has gradually been accumulating concerning the presence and function of stretch receptors in insects. The first description of muscle receptors (MRO) was made by Finlayson and Lowenstein (1955) in a study of Lepidoptera. This description was extended by Slifer and Finlayson (1956), Finlayson and Lowenstein (1958) and Osborne and Finlayson (1962), to cover in all twelve orders of insects. Among these orders only Neuroptera, Trichoptera and Lepidoptera possess muscle receptors which have an independent motor innervation. The sensory responses have been described for Odonata and Lepidoptera (Finlayson and Lowenstein, 1958), and for Lepidoptera in more detail (Lowenstein and Finlayson, 1960). These studies showed that the MRO's are "dual function" sense organs, signalling both tonic and phasic parameters of the stimulus. Finlayson and Lowenstein (1958) commented on the central nuclear region of the lepidopteran receptor and suggested that, since the receptor muscle (RM) was observed to be contractile, this region might be the analogue of the "nuclear bag" of the vertebrate muscle spindle. This resemblance has been restressed in more recent work on insect muscle receptors, and the time is clearly ripe for an investigation of the activities where the MRO's are important and of the ways in which they are used.

In common with many other "worm-like" insect larvae, the caterpillar moves in a peristaltic manner by lifting and promoting each segment in turn, starting at the posterior end (Kopec, 1919; von Holst, 1934; Barth, 1937). Such movements cannot avoid extending and releasing the MRO's, so these sense organs may very likely be involved in the co-ordination of locomotion. Segmental reflexes, such as those demonstrated by Pringle (1940), could be of importance in connection with a wide variety of activities. Consequently, after repeating and extending the observations of Lowenstein and Finlayson (1960) with a different experimental saline, the reflex responses to stretch of one or more MRO's were explored as widely as possible. (These results will only be summarized here in so far as they are relevant to the action of the MRO'S during locomotion, as it is intended to describe them in detail elsewhere.) Then experiments were performed on the acute neurophysiological preparation involving interference with the MRO's during wave-like activity, in order to shed some light on the role of these receptors during peristaltic locomotion.

II. Materials and Methods

Last instar larvae of *Antheraea pernyi* were used in all experiments concerned with the effects of stretching the MRO's and with the events during peristaltic activity. However, the motor axon innervating the RM could only be electrically stimulated in company with those innervating muscles of the dorsal body wall. Lifting the MRO with forceps almost invariably damaged several branches of the receptor motor nerve, so RM action had to be studied with the receptor *in situ*. Consequently, in the investigation of the effects of RM stimulation on the MRO sensory discharge, it was necessary to use pupae, as these could be immobilized more satisfactorily.

Most phytophagous insects have rather strange haemolymph inorganic cation concentrations, with a reversal of the usual Na/K ratio and a very high Mg concentration (Duchateau *et al.*, 1953). Thus, as indicated in the introduction, one of the first technical problems was to formulate salines based on the haemolymphs of the developmental stages used, which yielded electrophysiological responses similar to those seen in haemolymph. Details of the procedure used and the results obtained will be given elsewhere.

Preparations were opened by a mid-dorsal longitudinal incision, and pinned out along the edges of the cut to wax blocks on either side. Recording from nerves was achieved by using tapered platinum-iridium wire hooks, but in most cases efferent activity was recorded in muscles with intracellular microelectrodes similar to those described by Ballintijn (1961). The responses were led, either directly or through cathode followers, to conventional amplifying, display and recording apparatus. The MRO's could be lifted up in fine forceps mounted on a micromanipulator. The forceps were moved apart in a controlled manner by means of a mechanical waveform generator. Movement of the stretching forceps was signalled on the lower beam of the oscilloscope.

III. The Sensory and Reflex Background

When the MRO of the last instar caterpillar is stretched, the nature of the afferent discharge is considerably influenced by the medium in which it is immersed. In its own haemolymph and in the experimental media used here, this sense organ signals tonic and phasic stimulus parameters, as it does in Ephrussi and Beadle's (1936) *Drosophila* saline, the medium used in previous work on this receptor. However, while the total phasic discharge is not dissimilar in the above three media, the tonic discharge in the natural ionic environment is only two-thirds to one-half of the frequency at the same length in the *Drosophila* medium. Thus a re-investigation of the sensory responses with a variety of mechanical stimulating waveforms, including sinusoidal and constant velocity stretch, revealed that the phasic potentialities of the MRO are relatively greater *in vivo* than in salines with a more "normal" ionic content.

A proprioceptor consisting of a muscle strand with both viscous and elastic properties would be expected to take an appreciable time to become straightened when it was released. A well substantiated function of the intrafusal muscle fibres of the mammalian muscle spindle receptor is that of reflex length adjustment. The presence of a marked "silent" period in the sensory discharge from the lepidopteran MRO indicates that this is influenced by the viscous properties of the sense organ, and that as in the vertebrate receptor a useful function could be fulfilled by an RM system. The sensory discharge from both the mammalian (Leksell, 1945) and the decapod crustacean MRO's (Kuffler, 1954) is increased by stimulation of their RM's. The receptors of *Antheraea* pupae are similarly excited on stimulation of the receptor motor nerve when the strand is maintained isometric.

Eckert (1961) described in the crayfish a reflex resulting from stretch of an abdominal MRO which had the effect of inhibiting the afferent discharge both from the stretched receptor and from its neighbours up to three segments away. This effect was mediated via the sensory inhibitor nerve, a feature not present in insect MRO's (Finlayson and Lowenstein, 1960; and present observations). However, it remains possible that feedback reflexes affecting the caterpillar RM might result in sensory responses in the intact animal which were unlike those seen experimentally when the RM was inactivated. Experiments performed to check this possibility did reveal slight "autoinhibition" of the tonic discharge always seen in the RM when its central connections were intact, but this inhibition was phasic in nature with a relatively short time-constant of recovery. It was probably insufficient to result in more than a slight reduction in contrast between the sensory discharges in the stretched and unstretched states. Thus, the properties of the MRO with its central connections severed may be reasonably similar to its properties *in vivo*. This reflex probably serves to protect the sense organ from damage during intense stimulation and to "take up the slack" when it is released.

The classical work of Sherrington and co-workers (e.g. Liddell and Sherrington, 1924) showed the great importance of stretch reflexes in the maintenance of posture in mammals. More recent work, reviewed by Eldred (1960), has shown that these and other proprioceptive reflexes are also highly relevant to the co-ordination of locomotion. "Myotatic" reflexes like the stretch reflexes above have also been described in several invertebrate groups. Pringle (1940) described levator and depressor reflexes in the cockroach, and Bush (1962) showed the presence of similar reflexes in the limbs of decapod Crustacea. Proprioceptors situated where they can signal body position and locomotor movements commonly seem to mediate such negative feedback regulative reflexes. Recording from the body wall muscles of the caterpillar during stretch of one or more MRO's showed that this animal is no exception to the above general rule: the reflex effects were such as to resist extension and to "unload" the stretched receptor. Thus caterpillars possess proprioceptive equipment not unlike mammals, and this equipment activates at least one similar reflex function.

IV. The Role of the MRO's during Locomotion

Holst (1934) attached a thread to the posterior end of a caterpillar. If this was held firmly, any peristaltic locomotor wave present on the body stopped wherever it had arrived, and a new one started from the posterior end. He interpreted this occurrence as showing the presence of a proprioceptive reflex inhibiting the passage of the wave. The same effect was confirmed in the present study as occurring also in *Antheraea*, and it was further noted that an approximately isometric load on the posterior end was resisted with forces up to 80g peak tension. A resistance around 40g could be maintained by an intact caterpillar weighing only 8g, at a body length which precluded passive tension in the integument from being responsible (see Fig. 1). This last observation is strongly suggestive of a proprioceptive "resistance reflex", which could well be a manifestation in the intact animal of the stretch reflex described above. In view of the relatively high tension maintained between "steps", at a time when the observations of Barth (1937) and present results (see below) suggest that most of the muscles would normally not be contracting strongly, this reflex must involve a considerable proportion of the longitudinal body musculature. This response could therefore constitute an important element in the behaviour of crawling caterpillars.

One of the interesting features of the caterpillar is its hydrostatic skeleton. Almost certainly this feature, together with the method of locomotion, is what necessitates the very complex body musculature, involving about 80 independent motor units per segment. However, this feature also poses a very considerable technical problem: how to

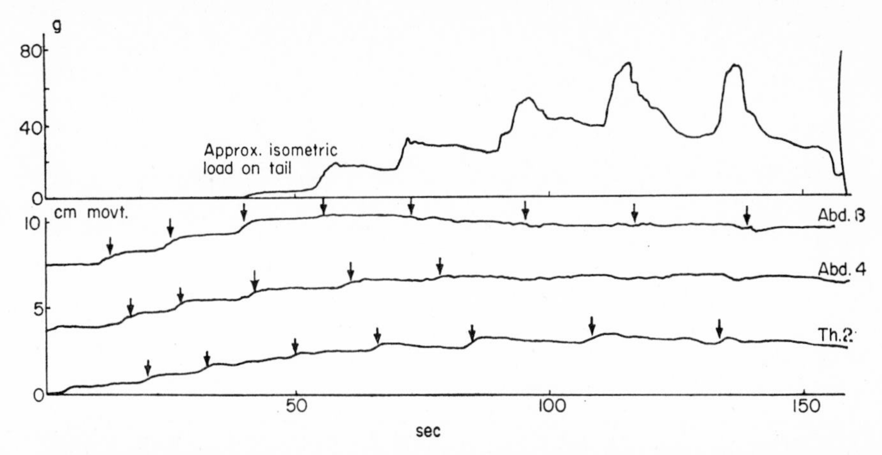

Fig. 1. "Resistance reflex" in a caterpillar. The figure is a tracing from a kymograph experiment where three isotonic levers and one isometric lever recorded the movements of an intact animal. The first three "steps" were unimpeded. The slower rhythm during loading was associated with slower wave propagation, and this together with the resistance constitutes the reflex.

interfere with the action of specific sense organs without incidentally producing other changes in the mechanical state of the animal which preclude locomotion. Holst (1934) was able to achieve this ideal while applying surgical techniques in a study of peristaltic locomotion, because the CNS and segmental nerves on which operations were performed are in some species visible through the integument. Comparable methods would be extremely difficult to apply to the MRO's owing to their small size, and all experiments here had to be performed on the acute preparation opened in the normal way. However, acute preparations of caterpillars which have started to spin their cocoons are much more excitable than preparations of feeding caterpillars and may show activity in the body musculature which resembles that described in intact crawling caterpillars by Barth (1937). This activity passes along the body in a wave-like manner and affects all the muscles which Barth considered to be important in peristaltic locomotion (see Fig. 2).

The only fact which renders the interpretation of these waves difficult is that they pass from the head posteriorly, rather than anteriorly, as they do when the intact caterpillar crawls forward. The movements of

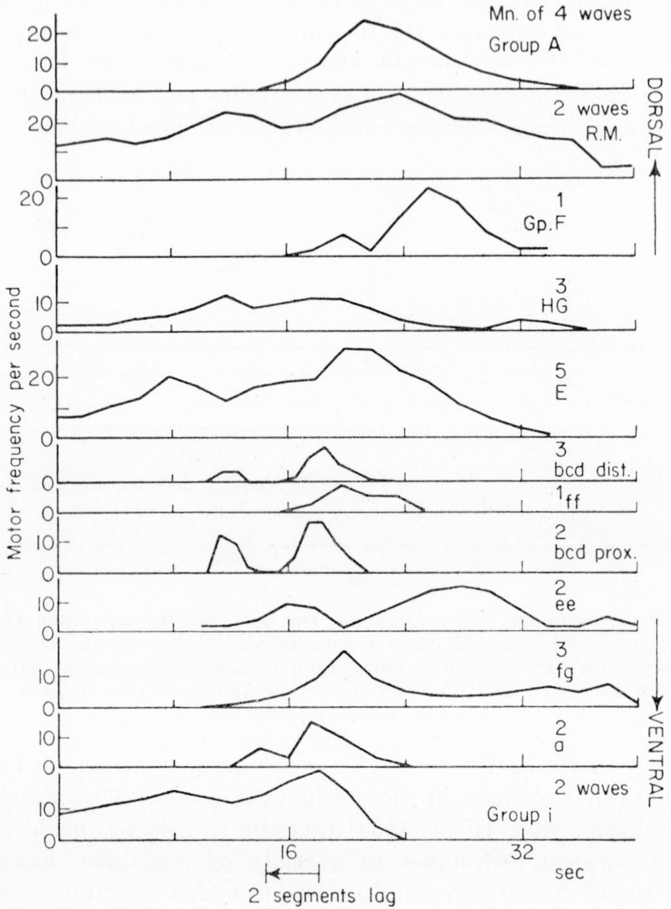

FIG. 2. Reverse peristalsis in an acute preparation of a spinning caterpillar (see text). The frequency of spikes is plotted on the ordinate for each muscle group listed on the right of the figure. The nomenclature is that of Lyonet (1762), slightly modified as appropriate to *Antheraea*. Groups A, E, bcd (dist. and prox.), a and i, are longitudinal muscles attached to the integument on or near to the intersegmental membranes. The remaining groups are long diagonal muscles forming antagonistic pairs: group F and group HG form one pair, lying close to the receptor (RM on the figure); groups fg and ee form the other, which lies ventrally.

thoracic legs and prolegs in such pinned-out caterpillars were such as would probably have moved the free animal backwards. Intact caterpillars do sometimes crawl backwards, however, and this action is achieved by a posteriorly directed peristaltic wave; this reverse crawling usually appears when the animal is in an aversive situation. The neurophysiological preparation is undoubtedly in a situation where it might be expected to exhibit reactions appropriate to noxious stimuli, so perhaps these movements are an expression of normal evasive activity. In that case, an investigation of this wave-like activity may shed some light on the co-ordination of normal forward locomotion.

Figure 2 is a composite graph compiled by recording from pairs of muscles during reverse peristalsis in the acute preparation. The muscles whose activity was recorded were chosen so that successive pairs had one member in common. The times of peak activity in the common member were then superimposed graphically for two pairs, thus giving the time relations of activity in three muscles. This process was then repeated for fifteen waves in four "crawling" preparations. The method is rather indirect and involves the assumption that successive waves travelled along the body at the same rates. In the few cases where measurements of wave velocity were made, this assumption appeared to be more or less justified, but it would clearly be preferable to use a larger number of recording electrodes if it were desirable to determine the precise time relations of activity in a large number of muscle groups. However, since the chief purpose of the present experiment was to show the qualitative nature only of these reverse waves, this method was thought adequate. The figure shows that the activity is complex and co-ordinated; furthermore, the dorsal and ventral longitudinal muscles which the observations of Barth (1937) implicated in normal peristaltic locomotion are active here in a manner which could produce in the intact animal a reverse locomotor wave.

While it is not possible in view of the reverse nature of these waves to be absolutely certain that they involve events similar to those during normal locomotion, it is of interest to speculate on the basis of these results concerning the way in which the RM is used during crawling. The muscle groups which lie near to the MRO and whose activity is shown in Fig. 2 are: group HG, group F, group E and group A. All of these groups are reflexly excited by MRO stretch and are so oriented that their contraction unloads the stretched receptor. Among these, group E and group A are certainly active earlier than the RM. (Two of the pairs of muscles from which recordings were made were group E with the RM, and group HG with the RM.) Even group A is maximally active before the second and larger peak of activity in the RM. Thus it would appear that the wave affects the body wall muscles in the vicinity of the MRO before RM activation, so that this will normally "take up the slack" in the released receptor as the segment shortens; but if the

MRO is held more or less isometric by an imposed load, then RM activation will result in MRO excitation and reflex reinforcement of the contraction of the dorsal longitudinal muscles. This is precisely what happened when the intact caterpillar encountered an isometric load, so the RM may act in the same way in the intact animal, playing a central part in the resistance reflex shown in Fig. 1.

The question of what part the MRO's play in "propagating" a wave of muscular activity along the animal is a more difficult one. Gray and Lissmann (1938) showed the presence of negative feedback stretch reflexes in the earthworm. They quoted the results of an experiment by Friedländer (1888), who showed that mechanical linkage alone was sufficient to maintain co-ordination between the two otherwise isolated halves of an animal crawling on a rough surface. Gray and Lissmann concluded that when the earthworm crawls on such a surface, successive activation of segmental stretch reflexes could be the major factor in conducting peristaltic waves along the body, though in the intact animal this would be reinforced by conduction along the nerve cord; on a smooth surface such nervous conduction would be more important. The musculature of the caterpillar and its movements during crawling are more complex than those of the earthworm. Nevertheless, in view of the very extensive reflex effects of MRO stretch, it is worth considering the possibility that a chain of segmental reflexes activated by these sense organs might alone be sufficient to co-ordinate a sort of locomotor wave, as in the case of the earthworm. Since there were some muscles active during locomotion which exhibited no reflex changes in activity on stretching a single MRO, it is unlikely that any such wave could resemble the pattern during peristaltic locomotion. Furthermore, the kind of reflex effects resulting from MRO stretch are very uniform, the majority being excitatory and occurring with more or less the same latency. However, since the reflex fields of adjacent receptors overlap, particular patterns of MRO excitation involving several receptors might have reflex effects not predictable by algebraic addition of the responses to stretch of each individual receptor. Finally, there might be special functions different in kind from those served by the stretch reflexes, functions only fulfilled by the MRO's during peristaltic locomotion. These considerations made it necessary to investigate the results of interfering with the MRO's during wave-like activity.

Accordingly, first one receptor was stretched during the passage of a wave, while a recording was being made from a muscle which in the inactive animal was reflexly excited by the same stimulus. The discharge was increased during the period of stretch, so the stretch reflex is confirmed to be active during peristalsis. It was thought possible

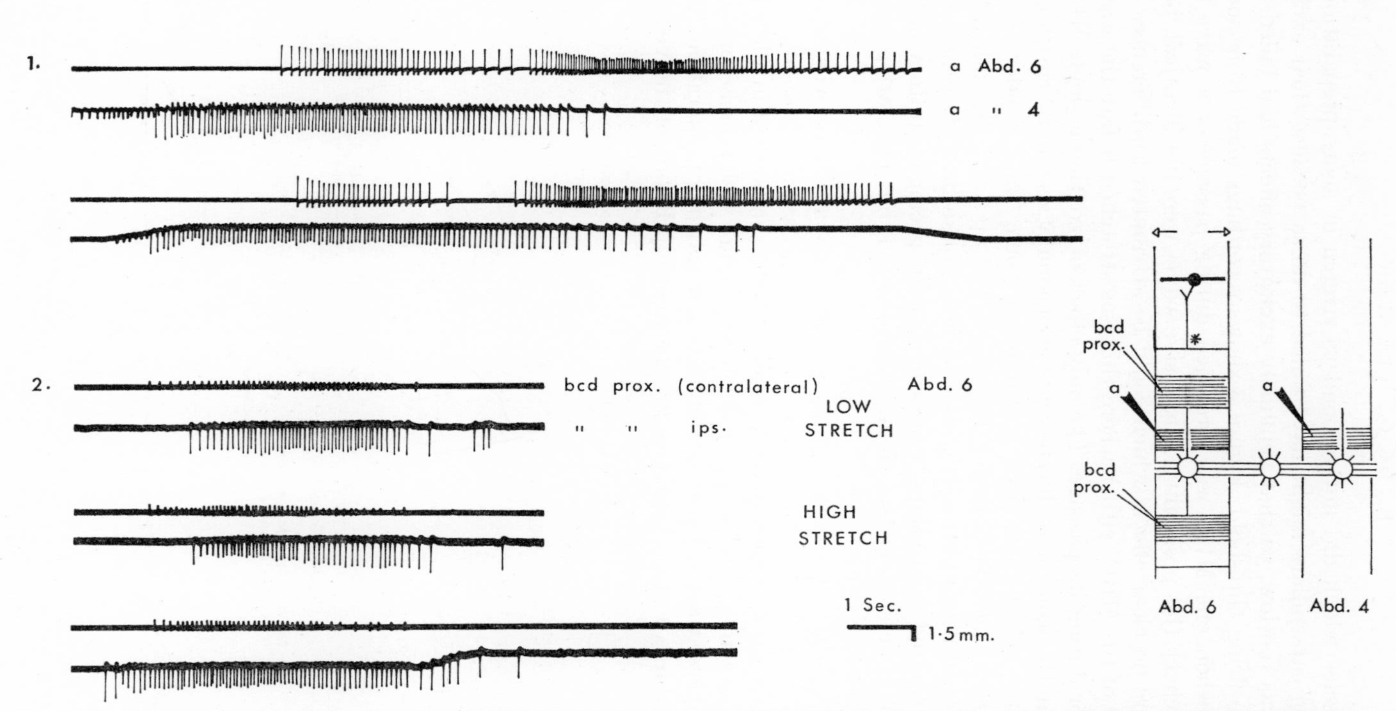

FIG. 3. Oscillograph recordings of wave-like activity in various ventral longitudinal muscles. The positions of these muscles are shown on the inset diagram. Fig. 3 (1) compares the rate at which a wave passes with and without concurrent stretch of a single MRO. Fig. 3 (2) consists of three records, each comparing activity in the same muscle on either side of one segment during wave passage. The first record was taken with the MRO on the left side of this segment unstretched and the second with the MRO stretched; in the third record stretch was applied towards the end of the wave. The horizontal calibration mark shows the time-scale, and the vertical mark the extent of stretch.

that muscles which did not respond to stretch in acute preparations of feeding caterpillars might have failed to do so because their excitability was too low, so that genuinely excitatory effects had failed to reach the firing threshold. Figure 3 shows recordings made from such muscles when one MRO was stretched during passage of a wave in order to check this possibility. Successive waves were not identical, but there were no clear effects which could be attributed to differences in the state of the MRO, either where this was stretched before the wave arrived or during its passage. The only effect of exciting a single MRO other than the usual stretch reflex was occasionally seen in a "crawling" preparation which had been inactive for some time. Under these circumstances, stretch was sometimes followed up to ten seconds later by antiperistaltic activity. The response was not sufficiently consistent to be attributed to anything more than a non-specific excitatory effect which could be achieved as often, or more often, by stimulating other sense organs.

Finally, MRO denervation was tested in order to see whether it produced any effects which could be interpreted as the result of interrupting a chain of segmental reflexes necessary or important in allowing the conduction of a wave along the body. Figure 4 shows the passage of two reverse waves along the abdomen of a crawling preparation. The upper graph is a plot of activity in muscle group P, one of the ventral

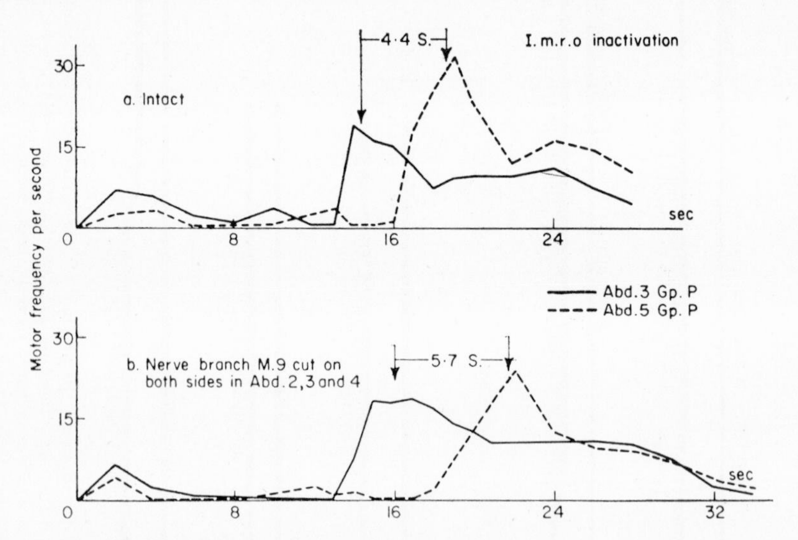

FIG. 4. The effect of MRO denervation in three segments on the passage of a reverse wave. The time taken for the wave to pass from abdominal segment 3 to segment 5 is shown on each graph. The wave was somewhat slower after denervation, but more data would be necessary to assess the significance of this effect.

diagonal groups in abdominal segments 3 and 5 with all sensory nerves intact. The lower graph shows the same thing after the purely sensory nerve, branch M. 9 in the nomenclature of Beckel (1958), had been cut in segments 2, 3 and 4 on both sides. This nerve contains the MRO sensory axon and other sensory fibres from the dorsal integument. Since a wave still passes under these conditions, the MRO's cannot be essential in any simple "chain reflex" manner to the conduction of waves of muscular activity of this kind along the body. This confirms the conclusion of Holst (1934) that wave conduction is in the caterpillar a primarily central nervous process. He found that section of all nerves in up to three segments did not prevent a wave from passing over the denervated region, and that after severing the insertions of all the locomotor muscles a minute degree of shortening still passed over the muscles remaining. By cutting single connectives on either side in such a way that the two hemisections were separated by two ganglia, Holst further confirmed that central nervous conduction played an essential part in this process. This method of cord section was necessary so that the part of the body posterior to the level of section should be sufficiently excitable to show peristaltic activity. Under these conditions the parts of the body anterior and posterior to the operated region showed independent crawling rhythms with no co-ordination between them. However, in view of the experiments of Gray and Lissmann (1946) on amphibian locomotion, extreme caution must be exercised in assessing the importance of the MRO's only on the basis of such experiments as have been performed here. Further study involving either total MRO inactivation in the otherwise intact animal, or else an investigation of integrative neuronal mechanisms in the CNS, would be required to shed light on this matter.

V. Conclusions

The MRO's of caterpillars activate a negative feedback stretch reflex in the muscles of the body wall whose chief purpose is probably to resist extension caused by changes in mechanical conditions at the periphery. While the muscular events of normal peristaltic locomotion may well be reinforced by segmental stretch reflexes activated successively along the body, such "chain reflexes" are unlikely to play more than a modifying part in the conduction of the peristaltic wave.

REFERENCES

Ballintijn, C. M. (1961). Fine tipped metal microelectrodes with glass insulation. *Experientia* **17**, 523–524.

Barth, R. (1937). Muskulatur und Bewegungsart der Raupen. *Zool. Jb.*, Abt. 2, **62**, 507–566.

Beckel, W. E. (1958). The morphology, histology and physiology of the spiracular regulatory apparatus of *Hyalophora cecropia*, L. (Lepidoptera). *Proc. 10th int. Congr. Ent.* **2**, 87–115.

Bush, B. M. H. (1962). Proprioceptive reflexes in the legs of *Carcinus maenas*, L. *J. exp. Biol.* **39**, 89–105.

Duchateau, G., Florkin, M., and Leclerq, J. (1953). Concentration des bases fixes, et types de composition de la base totale d'haemolymph des insects. *Arch. int. Physiol.* **57**, 149–162.

Eckert, R. O. (1961). Reflex relationships of the abdominal stretch receptors of the crayfish. I. Feedback inhibition of the receptor. *J. cell. comp Physiol.* **57**, 149–162.

Eldred, E. (1960). Posture and locomotion. *In* "Handbook of Physiology" (Field, ed.). Section 1, Vol. II, 1067–1088.

Ephrussi, B., and Beadle, G. W. (1936). A technique of transplantation for *Drosophila*. *Amer. Nat.* **70**, 218–225.

Finlayson, L. H., and Lowenstein, O. (1055). A proprioceptor in the body musculature of Lepidoptera. *Nature, Lond.* **176**, 1031.

Finlayson, L. H., and Lowenstein, O. (1958). The structure and function of abdominal stretch receptors in insects. *Proc. roy. Soc.*, B. **148**, 433–449.

Friedländer, B. (1888). Über das Kriechen der Regenwürmer. *Biol. Zbl.* **8**, 363–366.

Gray, J., and Lissmann, H. W. (1938). Locomotory reflexes in the earthworm. *J. exp. Biol.* **15**, 506–517.

Gray, J., and Lissmann, H. W. (1946). Further observations on the effect of deafferentation on the locomotory activity of amphibian limbs. *J. exp. Biol.* **23**, 121–132.

Holst, E. von (1934). Motorische und tonische Erregung und ihr Bahnenverlauf bei Lepidopterenlarven. *Z. vergl. Physiol.* **21**, 395–414.

Kopec, S. (1919). Lokalisationsversuche an zentralen Nervensystem der Raupen und Falter. *Zool. Jb.* Abt. 3, **36**, 453–502.

Kuffler, S. W. (1954). Mechanisms of activation and motor control of stretch receptors in lobster and crayfish. *J. Neurophysiol.* **17**, 558–574.

Leksell, L. (1945). The action potential and excitatory effects of the small ventral root fibres to skeletal muscle. *Acta physiol. scand.* **10**, Suppl. 31, 1–84.

Liddell, E. G. T., and Sherrington, C. S. (1924). Reflexes in response to stretch (myotatic reflexes). *Proc. roy. Soc.*, B, **96**, 212–242.

Lowenstein, O., and Finlayson, L. H. (1960). The response of the abdominal stretch receptor of an insect to phasic stimulation. *Comp. biochem. Physiol.* **1**, 56–61.

Lyonet, P. (1762). Traité Anatomique de la Chenille qui ronge le Bois de Saule. La Haye and Amsterdam.

Osborne, M. P., and Finlayson, L. H. (1962). The structure and topography of stretch receptors in representatives of seven orders of insects. *Quart. J. micr. Sci.* **103**, 227–242.

Pringle, J. W. S. (1940). The reflex mechanism of the insect leg. *J. exp. Biol.* **17**, 8–17.

Slifer, E. H., and Finlayson, L. H. (1956). Muscle receptor organs in grasshoppers and locusts (Orthoptera, Acrididae). *Quart. J. micr. Sci.* **97**, 617–620.

The Nervous Co-ordination of Insect Locomotion

DONALD M. WILSON

University of California, Berkeley, California, U.S.A.

The original work reported here was supported by NSF grant GB2116 and NIH grant NB3927. I am indebted to R. J. Wyman for assistance with some of the computations and for permission to quote his recent work on dipteran flight control.

It is now well established that for some types of rhythmic behaviour the basic pattern of motor output can be generated by a central nervous system which is isolated from sources of phasing information. This has been shown for a flying insect (Wilson, 1961), and it might seem a reasonable guess that this would also be true for walking insects, but experiments do not prove this. It is also known that reflexes play a role in locomotion. The purpose of this paper will be to suggest how peripheral reflexes may influence the operation of a central pattern generator.

Two reflexes will be discussed. First, there is the stretch reflex which controls wingbeat frequency in locusts and which operates with such slow time constants that cycle-by-cycle influence is lost. Secondly, there is the proprioceptive leg reflex which presumably influences leg posture during walking and which shows precise phasic influence, even during the most rapid leg movements.

These two reflexes may represent extremes on the spectrum of relative speeds of operation. An already known intermediate case is that described by Weevers (1965) for caterpillars. A second slow reflex operates in locust flight and it regulates forewing twisting and lift (Gettrup, 1964).

I. Locust Flight: the Central Oscillator

During normal flight the individual motor units fire once or twice per wingstroke. The different muscle groups (elevators and depressors of the two wing pairs) are activated in a precise sequence which has already been described in considerable detail (Wilson and Weis-Fogh, 1962). The basic elements of this pattern are present in the discharge in the motor nerves even after removal of all wing sense organs and muscles.

The behaviour of a single unit during flight is illustrated in Fig. 1. In (a) the electrical record shows single or double activations during each wingstroke (the time marker is at 10 c/s). In (b) the consecutive intervals from about one minute of steady flight are recorded in a

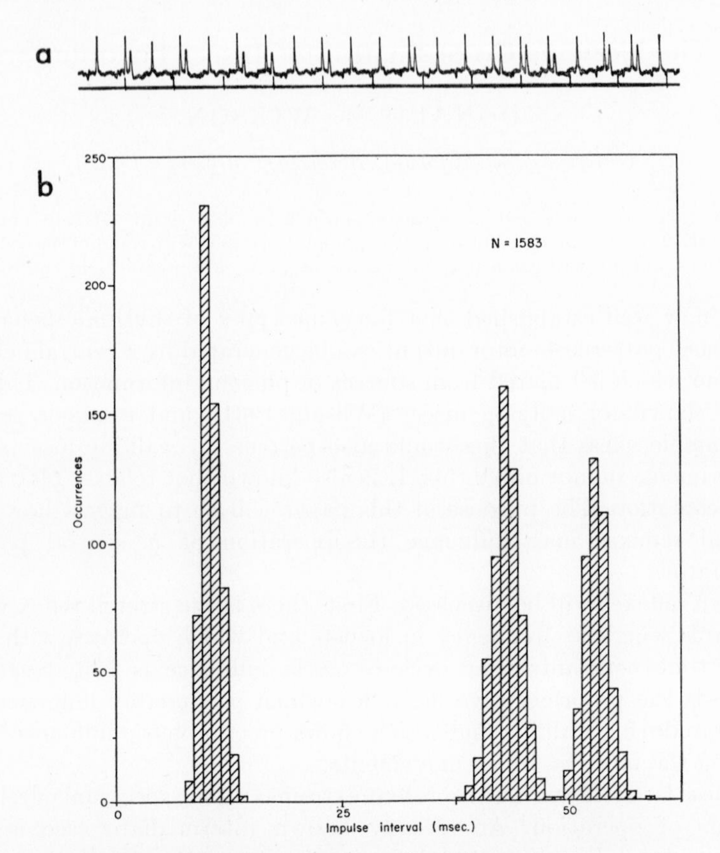

FIG. 1. *Top.* Record from a flight muscle of a locust during normal flight. The unit is just at threshold for firing twice per cycle. The second muscle action potential is small due to antifacilitation or relative refractoriness of response. The time mark is at 10 c/s. *Bottom.* Histogram of intervals from approximately one minute of flight. For explanation see text.

histogram. The left-hand mode includes the intervals between pairs of impulses, and the extreme right-hand one the wingbeat intervals when there was only a single impulse during the preceding cycle. The intermediate mode includes intervals from a pair to the next wingbeat. In different muscles, different flights or different specimens the areas of the three histogram modes vary (there is variation in the amount of multiple firing), and the modal values vary (there is variation in wingbeat frequency). When less steady flights are plotted the variation results in fusion of the two longer interval modes.

If the behaviour of the single units is observed at different temperatures, it is seen that only the interval between the paired impulses shows a strong temperature dependence. The modal value of about 10 msec at 26° C is reduced to about 6 msec at 32° C. The wingbeat interval changes only slightly (Weis-Fogh, 1956). This low temperature dependence of the wingbeat cycle duration prevails even after surgical removal of the stretch receptors (see later) or total removal of the wing phased sense organs. The low temperature dependence is therefore a property of the inherent central oscillator and not a consequence of mechanical effects measured by the proprioceptors.

The temperature dependence of the interval between the paired impulses is similar to that expected if the pair were generated by refractory oscillation during a single long synaptic potential or other state of excitation. Other evidence suggests that the paired discharge is due to relaxation oscillation with the interval determined by relative refractoriness (Wilson, 1964). The relative refractoriness of the axon, at least, is quite long (easily detectable for 30–50 msec). Antidromic impulses can cause resetting of the discharge timing for about 10 msec after the antidromic spike. During flight this resetting can be effective within the single wingstroke burst of impulses, but antidromic impulses have no effect on subsequent cycles of activity or on other muscles during the same cycle. It seems reasonable to conclude that relative refractoriness of the motorneurone is responsible for this one parameter of the motor output pattern during flight, namely, the length of the interval between the impulse pairs (or multiple bursts) during a single phase of the wingstroke. It does not apparently affect any other aspects of the pattern. For these we must seek other mechanisms.

Random stimulation of the nerve cord of deafferented preparations can elicit the normal flight pattern of motor output, but at reduced frequency. In anoxic or otherwise poor preparations it sometimes happens that only one or two motor units of the flight system will respond to stimulation of the nerve cord. When this happens, the individual units do not show the normal behaviour of Fig. 1, but instead produce a sequence of intervals with a single modal value and a large scatter. Figure 2 shows a histogram of intervals from such a unit responding to random input. The dead time at the beginning is due to refractoriness. The gradual increase to the peak value indicates some kind of counting which could be due to either facilitation or accumulating refractoriness. The declining tail approximates the exponential distribution of the input. Since there are no humps in this distribution it is to be concluded that the unit, under these conditions at least, had no tendency toward autorhythmicity other than that which is imposed by refractoriness. For a record such as that accumulated in the histogram of Fig. 2, the serial correlation coefficient is significantly positive for the first three to four orders, indicating that even during a random input sequence the state of excitation of the cell waxes and wanes relatively slowly as compared with the fluctuations in the input.

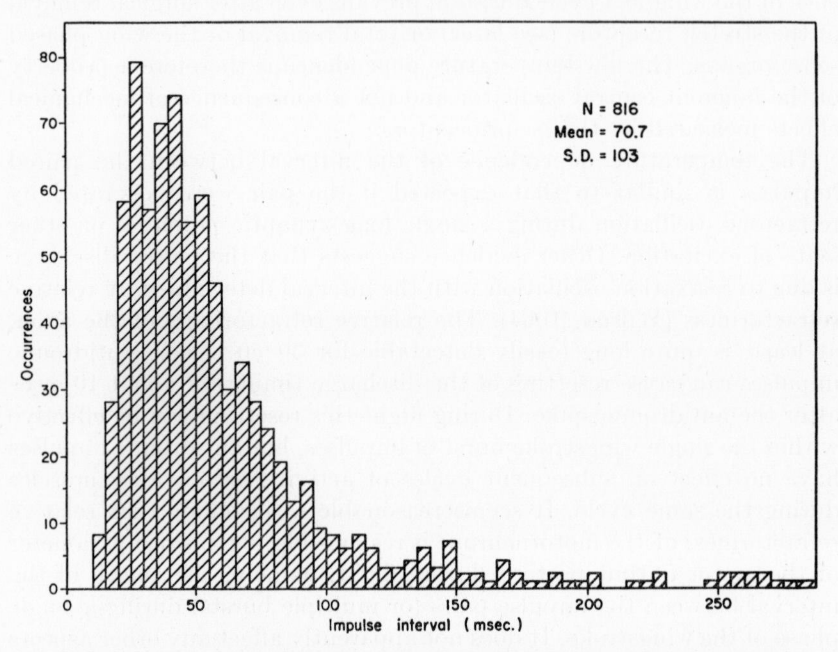

FIG. 2. Interval histogram for response of only a single unit of the flight system during random interval stimulation of the abdominal nerve cord. Only a single mode of response is present.

When two units in the same muscle respond to random stimulation of the nerve cord they may behave quite independently, even though in flight they show perfect synchrony. Figures 3(a) and (b) give spike interval histograms for two units of a tergosternal muscle which is used only in flight. Unit A had the higher frequency, the greater rhythmicity (less scatter in histogram), and the most pronounced frequency trends (as shown by the serial correlation coefficient (d)). This combination of properties seems to be common in the case of relatively more excited neurons. However, both units showed short frequency trends which were nearly synchronous and in the same direction (i.e. positively correlated). These units were rather surprisingly found to be phasically independent. In Fig. 3(c) the fractional time of occurrence of A spikes in B intervals is plotted showing that there is practically no preference for particular times. During the operation of only two synergistic units of the flight system no special element of the normal pattern can be detected.

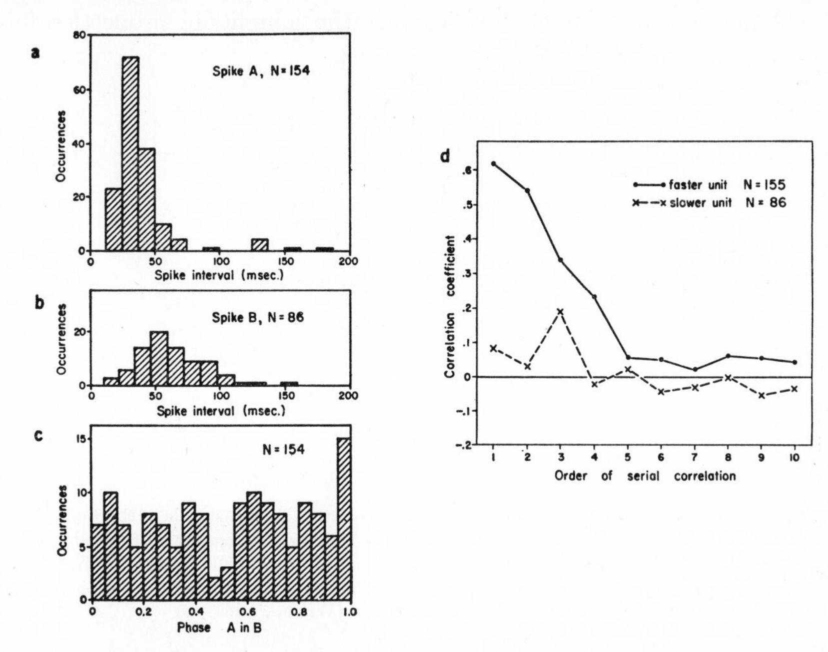

FIG. 3. Response of two units in the same muscle during random interval stimulation of the nerve cord. (a) and (b) are interval histograms for the two muscle action potentials, (c) is the phase histogram, and (d) shows the first 10 orders of serial correlation coefficient. The higher frequency unit has the least scatter and the more pronounced brief frequency trends. Although the two units were highly correlated in frequency the phase relationship is nearly random. (After Wilson and Wyman, 1964.)

Normal behaviour of the flight motorneurones seems to require the activity of antagonistic groups. It is probable that the normal oscillation never occurs without many units being active, but for convenience one may think about a single reciprocating pair rather than two populations of cells. An artificial neurone network, involving reciprocal inhibition, has been described which can mimic the activity of the flight neurones accurately (Reiss, 1962; Wilson, 1964). This model depends upon interneuronal elements which have not been detected experimentally (see Fig. 4). Pavlidis (1964) has developed an alternative analogue and mathematical model which does not include interneurones, but does require either (1) a long lasting autoinhibition combined with the reciprocal inhibition, or (2) cross excitation operating through a long lasting delay. Either (1) or (2) is required in order that one train of impulses may stop long before the antagonistic one begins.

With several simple electronic models of the flight system described it should now be possible to examine the ganglionic properties for

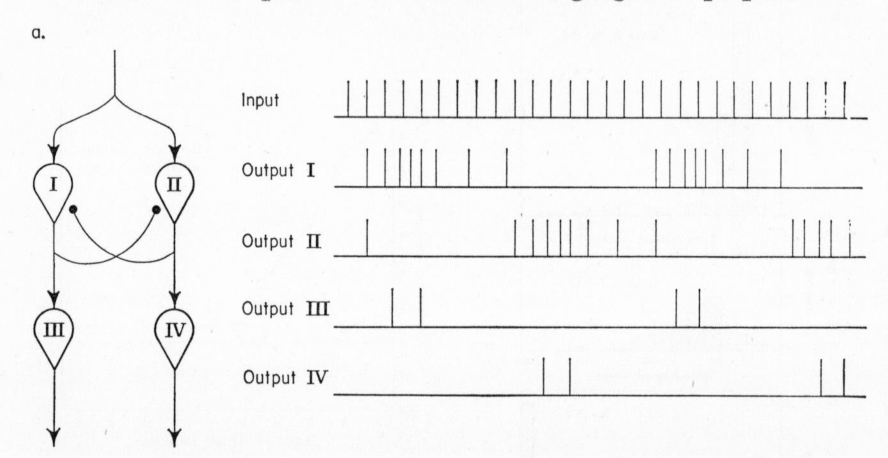

Fig. 4. A hypothetical reciprocal inhibition network which can simulate the behaviour of locust flight motorneurones. The input may be regular or irregular without changing effect on the output. Cross inhibiting units I and II produce the basic antagonistic pattern. Follower neurones III and IV produce the motor output with long periods of silence between bursts. (From Wilson, 1964.)

relevant details. For the present we can only assume that some oscillation exists in the ganglion and that it consists in its simplest form of a network of neurones.

II. Reflex Modulation of the Locust Flight Central Control System

It was mentioned above that the deafferented locust flight control system operates at reduced frequency. It was found by Wilson and Gettrup (1963) that a set of stretch receptors, one at the base of each wing, are necessary for the maintenance of the normal wingbeat frequency, and that no other phasically patterned sensory input contributes significantly to this function. Further work (Wilson and Wyman, 1964) has been directed towards ascertaining the extent to which these stretch receptors supply the CNS with necessary phasic information about the wing movements. It is known that the sensory discharge itself contains information on phase, frequency and amplitude of the wingbeat (Gettrup, 1963). However, it can be shown that little of this information is preserved in the central processing. Rather, it appears that the impulses from the stretch receptors lose their phasic information content by a process analogous to integration by a long decay constant RC filter, and that it is the resulting nearly DC level of

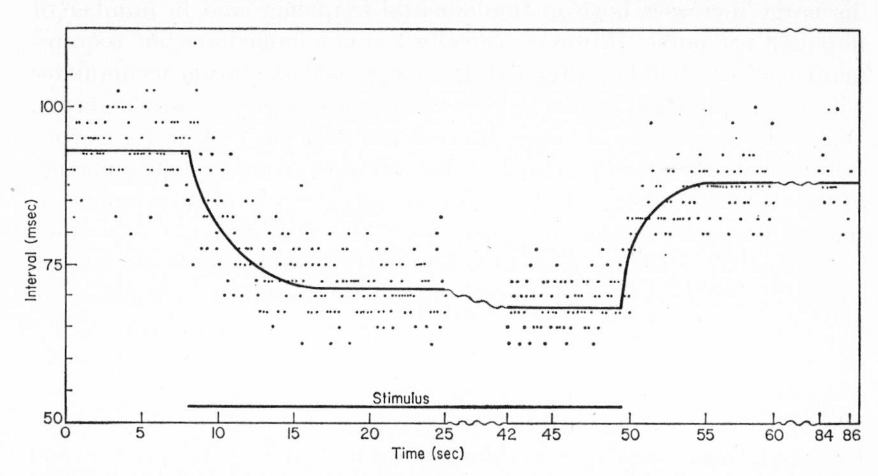

FIG. 5(a). Response interval (= wingbeat interval) vs. time in a wingless locust preparation which produces the flight response due to the wind-on-the-head stimulus. The frequency is increased during electrical stimulation of proximal stumps of two of the four stretch receptor nerves. Each dot is one interval. The line is a visual approximation of the average. The response takes many cycles to wax and wane.

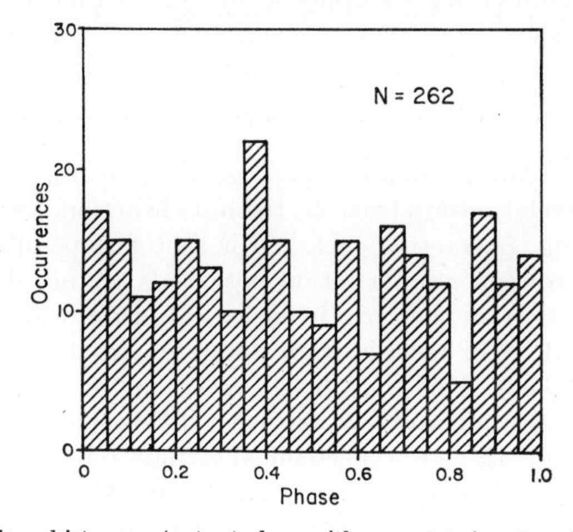

FIG. 5(b). The phase histogram (output phase with respect to input cycle) for part of the same experiment as in (a). (a and b after Wilson and Wyman, 1964.)

excitation which affects the frequency of the central oscillator. The evidence for this is the following:

If the thoracic sense organs are ablated and the muscle response recorded during the normal flight stimulus of wind-on-the-head, a weak, slow, but normally patterned discharge is seen. If now two of the stretch receptor proximal nerve stumps are stimulated, the muscle discharge increases both in fundamental frequency and in number of impulses per burst. However, the effect is not immediate but requires many cycles to build up (Fig. 5a). It appears as if excitation accumulates with a time constant of one to two seconds or several tens of wingbeats. When the stimulation is discontinued the frequency of motor output falls with a similarly long time course. Even during the steady portion of the response (when the filter is saturated) there is no entrainment or tendency towards phase locking (Fig. 5b). It is this fact especially which suggests that a DC component intervenes between input and output oscillators. Otherwise, an oscillator which affects the frequency of another oscillator would also affect its phase. Since the input to our hypothesized filter is a pulsed one, we can expect that the output of the filter will include some ripple and that this ripple should have some effect on the motor pattern. However, in the neatest experiments the ripple is too small to be detected even by averaging over several hundred cycles of output.

The locust stretch reflex is a tonic one, even though it controls the frequency of a very rapid rhythmic activity. Its input is averaged over many wingbeat cycles, and the resultant state of excitation modulates the frequency, and perhaps amplitude, of the central oscillator without affecting its phase.

III. Control of Dipteran Flight Muscles

In insects such as flies which have myogenic flight muscles, there is no fixed temporal relationship between the muscle action potential and the several following contraction cycles. It is therefore possible that phase relationships between muscle action potentials are not determined by some fixed pattern as they are in locusts. Indeed, this lack of phase coupling is evident in records from different muscles in flies (Wilson and Wyman, 1963). This may be observed in the records in Fig. 6. In this figure lines one and two are from different dorsal longitudinal muscles, while line three is a mechanical artifact representing the wing-beat. If one compares, for example, the largest units in lines one and two, then it can be seen that the two differ in frequency and drift past each other in phase. A phase histogram from a long record shows no

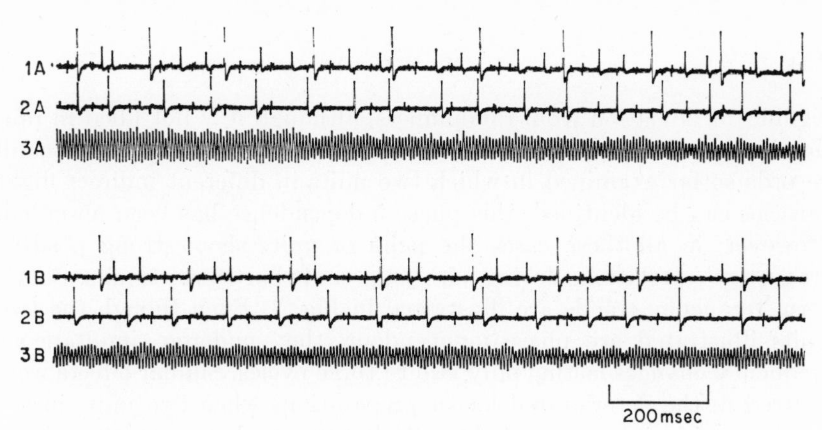

FIG. 6. Records from the dorsal longitudinal muscle motor units of a flying fly. Lines 1 AB show at least three units from the same muscle. The three units can show a pattern for many cycles. The units in 2 AB from the muscle of the other side show no relationship to those of 1 AB. The lower lines record a mechanical artifact showing the wingbeat frequency. (After Wyman, 1964.)

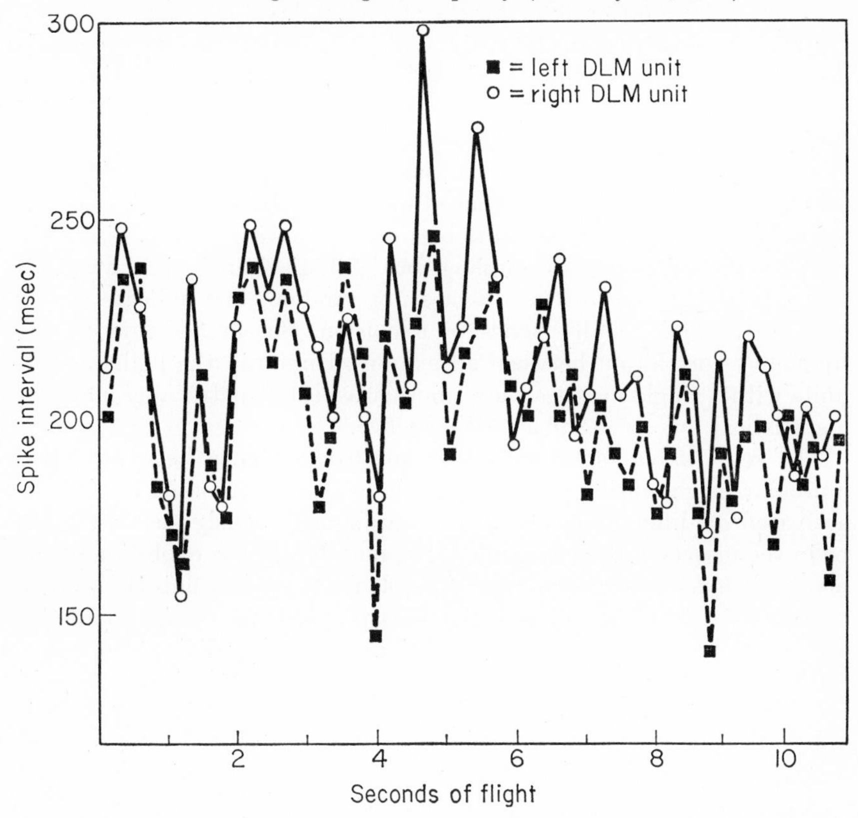

FIG. 7. Plot of spike interval vs. time for two dorsal longitudinal muscle units in a flying fly. Short term trends lasting only two to four cycles are evident in this record. Both units show the trends, and the two are positively correlated. Nevertheless, the two units showed no significant phase dependence. (After Wyman, 1964.)

significant deviation from randomness, although it is not filled in randomly (i.e. the phase shifts gradually without much jitter). In all records so far examined in which two units in different indirect flight muscles can be identified, this phase independence has been observed. However, in all these cases the pairs of units show strong positive frequency correlations during both long- and short-term trends. Short-term frequency trends are illustrated in Fig. 7. Even though the two units illustrated are phase independent, they undergo simultaneous frequency changes lasting only two to three cycles. Similar effects were noticed in the deteriorated locust preparations when two units in the same muscle were compared. This result suggests that we must envisage a neural model in which spike initiation is not closely coupled to the arrival of presynaptic spikes, but rather is dependent upon more smoothly changing excitatory state. This is reminiscent of the requirement of an integrating filter for the stretch reflex.

In the records 1A and 1B of Fig. 6, three units are easily recognized. These units are all within one dorsal longitudinal muscle. If these are called *l* (large), *m* (medium) and *s* (small), then the pattern in 1A is *l*, *m*, *s*, *l*, . . . etc. In 1B it begins with the only other possible permutation *l*, *s*, *m*, *l*. . . . but then in the third cycle *s* and *m* are synchronous and the pattern flips back to that of line 1A. These transitory patterns could be due to slow phase shifts between units with slightly dissimilar frequencies and no phase coupling, but statistical analysis shows that they are not. Phase histograms are not random, but indicate strong preferences and usually zones of exclusion. So, in addition to the frequency correlation between all examined units in the flight motor, units within single muscles show another aspect of patterning, namely, a statistical but strong phase dependence.

The remarkable fact about the phase dependence between intramuscular neighbouring units is that these synergistic units tend to be activated at different times rather than synchronously, as they must be in the insects with neurogenic flight muscles. In the evolution of the myogenic mechanism it has been possible for the motor units to become asynchronized by the degeneration of a previous mechanism, and this appears to be the case for the populations controlling different muscles. However, the development of a new special phase relationship would seem to require some adaptive significance. Although the argument is not perfect, it seems probable that the advantage is the following. After a single nervous activation ordinary fast muscles produce a twitch of well-defined temporal characteristics. The myogenic muscles, when properly loaded, oscillate several times instead, but the envelope of the oscillation has a time course similar to a twitch in that there is a gradual falling phase. In order to maintain a steady amplitude of oscillation, nerve impulses must arrive more often than some minimal frequency. This frequency is about 10 c/s in the case of the beetle *Oryctes* (Machin and Pringle, 1959). This value is unknown for fly muscles, so that there is some uncertainty in this hypothesis. During flight the motor units of flies may be activated only 3 or 4 times per second in weak performances. This is probably insufficient to maintain a smooth power output from the single units, but by staggering in time the various synergistic units of the same muscle, a smoothing will be achieved. This is presumably the purpose of the considerable neurophysiological rearrangement between the locust type mechanism and that of the fly.

It is too early to attempt to describe the variety of patterns of motor control of flight in the different varieties of insects, but it is possible that a rich comparative neurophysiology will be revealed in the future. Analysis of the co-ordination of the flight muscles of a bug (*Lethocerus*) has shown basic similarity to the flies (Barber and Pringle, 1964). This comparative approach may have value for understanding the individual cases. Because of the extensive anatomical homology between the flight parts, the presumed phylogenetic relationships between the various groups, and the basic similarity of nervous structure, I feel that a model of the integrative mechanisms for one insect should be convertible to that of another by parameter adjustment rather than by inclusion of different types of functions. This criterion limits the range of models which can be constructed for each case.

Another point of interest in regard to the evolution of the more special myogenic flight patterns involves the reflex effects due to sensory input from the wings. In locusts, this input is timed by the movements of the wings. This is almost certain to be found true also in the myogenic rhythm cases (at least the low frequency ones such as *Lethocerus*). However, in these animals the motor output is not phase locked to wing movements. Therefore, the input–output relationship is necessarily phase independent. In the neurogenic case both input and output are synchronized with the wing movements and therefore show a phase correlation; however, experimental manipulation has shown that this correlation is unnecessary for correct operation of the ganglionic transfer function. In this respect the locust shows an appropriate sort of pre-adaption for the evolution of myogenicity.

As to the motor patterns themselves, the fly may be considered a degenerate locust; it has lost most of the phase locking between motor units and between nervous and mechanical cycles. In both fly and locust it is probable that frequency modulations in the pulse trains controlling the muscles accomplish changes in power and spatial orientation. In the locust, this frequency modulation is superimposed upon a pattern of oscillatory interaction between reciprocating antagonists. In the fly the control system approaches the ideal pulse frequency modulation code, in which the timing of the individual pulse has no importance.

IV. Cockroach and Grasshopper Leg Reflex

Reflexes which show phase independence between input and output may be called tonic reflexes. Tonic reflexes are appropriate for postural adjustment and for control of rhythmic activities in which it is permissible that the mechanical output be averaged over several cycles. This may be true of such things as visceral regulatory reflexes and locomotion in smooth media like air or water. It could also be true of any very high-frequency rhythm, especially one involving several effectors operating in parallel, for example, the several legs of a running insect. From some cinematograph records of running cockroaches (*Blata orientalis*) I have determined that these animals can move their legs as fast as locusts beat their wings. If the proprioceptive leg reflexes were similar in temporal properties to those of the wings, then these reflexes could not account for the co-ordination of the legs, but could still produce adaptive adjustments of gait over several cycles by tonic or postural effects.

An experiment was designed to test the frequency capabilities of the proprioceptive leg reflexes. An entire leg of an intact animal was manipulated and records made of the activity of certain motor units, usually in the extensor of the tibia. The leg was attached at the tarsus to a lever driven by a powerful galvanometer. The galvanometer was controlled by an electronic function generator and the movements of the lever monitored by recording the area of an interrupted light beam. Both grasshoppers and cockroaches were examined, usually the mesothoracic legs of the former and metathoracic legs of the latter. Results were generally similar for the two types of animal, even though their walking frequencies are quite different. The legs, of course, contain many proprioceptors (campaniform sensillae, hair sensillae, and chordotonal organs), and it is not known which are effective in the response studied.

With periodic leg movements the motor units respond in a phasic manner at all frequencies studied. With low amplitude stimulation small muscle potentials (presumably slow units) result, and with more vigorous stimulation the large fast fibre response is superimposed. The fast and slow units of a muscle appear to have the same response characteristics but different thresholds. There is a good response to the first cycle of stimulation, and the response may continue for hundreds of cycles. At other times fatigue is rapid, and the output not only decreases in volume but also tends to fuse and lose its phasic aspect.

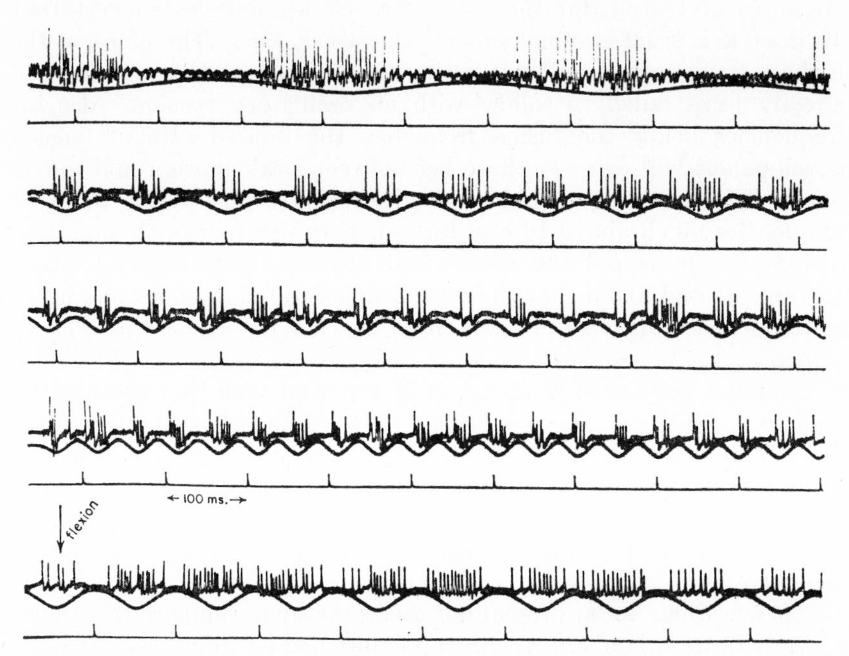

FIG. 8. Response of a fast motor unit in the extensor of the tibia of the hindleg of a cockroach (*Periplaneta*) during constant amplitude (*ca.* 6 mm) variable frequency movement of the whole leg approximately in the normal running plane. Top line, muscle potentials; middle line, record of movement; lower line, 10 marks per second. The records are at about 3, 10, 15, 20 and 10 cycles per second. The variable amplitude of the action potentials is due to antifacilitation or refractoriness during high-frequency trains. The beginning of a stimulus bout may be seen in the fourth record (at 20 c/s). The response begins within the first quarter of a cycle.

Figure 8 illustrates some results of nearly sinusoidal stimulation. A single fast unit is recorded. The amplitude of the spike decreases with increased frequency of response (it antifacilitates). With stimuli between 3 and 20 c/s a nicely phasic response is maintained without noticeable lag. In some cases the system has been driven to respond at up to 30 and 40 times per second or more than twice the speed at which it is ever used by the animal.

Measurements of latency from the beginning of the first cycle of motion show that the minimum reflex time is no more than 10 msec. This includes the activation of the sense organs, centripetal conduction, ganglionic delay, centrifugal conduction and neuromuscular delay.

It appears from this that these insects have their running in no way limited by the speed of operation of their leg proprioceptive reflexes. A significant limit must be the rate of muscle contraction and relaxation. Measures of twitch duration of cockroach leg muscles suggest that 40 msec is a usual minimal value (Usherwood, 1962). Therefore, at the highest frequencies studied for the nervous reflex, the muscles will already have failed to follow with an oscillatory tension. Also, at frequencies below the fusion frequency the limited rate of tension development will cause a phase lag between peak tension and movement. This phase lag should reduce the effectiveness of the reflex at frequencies much above 10 c/s. Indeed, Horridge (personal communication) has measured the mechanical response of the leg to forced movements and found that the system failed to follow sinusoidal input at frequencies much lower than those which I report for the nervous response.

The reflex response recorded here is arranged such that the muscles tend to resist length change. Pringle (1940) found the same thing due to step function inputs when recording from the same leg and the reciprocal relationship in the contralateral leg. This crossed reciprocal reflex was found in the present experiments also. It works at rather high speeds, but the minimal time found for the crossed reflex was about 25 msec.

It is tempting, but at present not useful, to try to compare the results of study of the whole reflex with those obtained on single sense organs. Pringle and Wilson (1952) and Chapman and Smith (1963) have studied dynamic responses of the sensory spines of cockroach legs in order to describe the transfer function of this single stage of nervous processing. Perhaps the ultimate aim of this kind of study is to be able to fit together the several serial functions which result in the total operation of the reflex. Altogether this will be a massive undertaking since several populations of different kinds of sense organs, each with different individual functions, may operate in parallel and then converge with different weights of effectiveness in the ganglion. In order to gain the quickest understanding of the reflexes, it may be most profitable to study them as wholes and hope that we can fit these into a mechanistic description of the relatively complex phenomenon, locomotion.

V. Summary

1. The flight motor of locusts is under the control of a central nervous pattern generator. The frequency of oscillation of the central rhythm is controlled by a reflex involving stretch receptors at the base of the

wings. The ganglionic input and output of the reflex are phase independent.
2. The central oscillator appears to consist of reciprocal interactions of antagonistic motor units, since the individually active units do not show elements of the flight pattern.
3. In flies the motor output is independent in phase with respect to the wing movements, and activations of the various muscles are also temporally independent. For the most part, control is by simple frequency modulation. However, units within one muscle are phase locked in a staggered sequence, apparently insuring smooth mechanical effects of the whole muscle.
4. Proprioceptive reflexes in cockroach and grasshopper legs are much faster than those of the wings and can transmit phasic information at more than 30 c/s during sinusoidal mechanical input. The limiting factor in the ability to adjust individual steps reflexively may lie in the muscle capabilities rather than in the nervous phenomena.

REFERENCES

Barber, S. B., and Pringle, J. W. S. (1964). The functional organization of the flight system in belostomatid bugs (Heteroptera). *XII Int. Congr. Ent.* (In press.)

Chapman, K. M., and Smith, R. S. (1963). A linear transfer function underlying impulse frequency modulation in a cockroach mechanoreceptor. *Nature, Lond.* **197**, 699–700.

Gettrup, E. (1963). Phasic stimulation of a thoracic stretch receptor in locusts. *J. exp. Biol.* **40**, 323–333.

Gettrup, E. (1964). Control of fore-wing twisting by hind-wing receptors in flying insects. *XII Int. Congr. Ent.* (In press.)

Machin, K. E., and Pringle, J. W. S. (1959). The physiology of insect fibrillar muscle. II. Mechanical properties of a beetle flight muscle. *Proc. roy. Soc.* B. **151**, 204–225.

Pavlidis, T. (1964). Analysis and synthesis of pulse frequency modulation feedback systems. Report no. 64–13, Electronics Research Lab., Univ. of Calif., Berkeley. 150 pp.

Pringle, J. W. S. (1940). The reflex mechanism of the insect leg. *J. exp. Biol.* **17**, 8–17.

Pringle, J. W. S., and Wilson, V. J. (1952). The response of a sense organ to a harmonic stimulus. *J. exp. Biol.* **29**, 220–234.

Reiss, R. F. (1962). A theory and simulation of rhythmic behaviour due to reciprocal inhibition in small nerve nets. Amer. Fed. of Information Processing Societies, Proceedings Spring Joint Computer Conference, **21**, 171–94.

Usherwood, P. N. R. (1962). The nature of "slow" and "fast" contractions in the coxal muscles of the cockroach. *J. Insect Physiol.* **8**, 31–52.

Weevers, R. de G. (1965). Proprioceptive reflexes and the co-ordination of locomotion in the caterpillar of *Antherea pernyi* (Lepidoptera). *In* "The Physiology of the Insect Central Nervous System" (J. E. Treherne and J. W. L. Beament, eds.), pp. 113–124, Academic Press, London and New York.

Weis-Fogh, T. (1956). Biology and physics of locust flight. Part II. Flight performance of the desert locust (*Schistocerca gregaria*). *Phil. Trans. roy. Soc.* B. **239**, 459–510.

Wilson, D. M. (1961). The central nervous control of flight in a locust. *J. exp. Biol.* **38**, 471–490.

Wilson, D. M. (1964). Relative refractoriness and patterned discharge of locust flight motor neurons. *J. exp. Biol.* **41**, 191–205.

Wilson, D. M., and Gettrup, E. (1963). A stretch reflex controlling wing-beat frequency in grasshoppers. *J. exp. Biol.* **40**, 171–185.

Wilson, D. M., and Weis-Fogh, T. (1962). Patterned activity in co-ordinated motor units, studied in flying locusts. *J. exp. Biol.* **39**, 643–667.

Wilson, D. M., and Wyman, R. J. (1963). Phasically unpatterned nervous control of dipteran flight. *J. Insect Physiol.* **9**, 859–865.

Wilson, D. M., and Wyman, R. J. (1964). Motor output patterns during random and rhythmic stimulation of locust thoracic ganglia. *Biophys. J.* (In press.)

Wyman, R. J. (1964). Probabilistic characterization of simultaneous nerve impulse sequences. (In manuscript.)

The Central Nervous Control of Respiratory Movements

P. L. MILLER

Department of Zoology, University of Oxford

I. Introduction

The examination of a simple motor output from the central nervous system would seem at first sight to provide a promising starting point for a study of some aspects of the underlying organization of neurones. Spiracle movements are simple, and they are produced by motor impulses which arise endogenously in the ganglia. Moreover, the trains of impulses often remain relatively unperturbed when parts of the nerve cord are sectioned or otherwise manhandled, and they continue when more complex patterns of activity have disappeared in ageing preparations. On the other hand, the conclusions reached may not apply to other motor systems since the median nerve, which innervates the spiracles, is thought to belong to a separate "autonomic" nervous system, and may fuse with the remainder of the cord late in embryological development. Its functional independence may therefore represent development different from that of the rest of the ganglion. With these hazards in mind, however, something of the co-operation between motor cells and interneurones in producing the patterns of motor impulses to the spiracles can be learnt.

The spiracle mechanisms can be divided conveniently into a "power" system responsible for shutting the valve and a "control" system which regulates the amount of opening of the valve when power is reduced or switched off. I shall discuss these aspects in two contrasting spiracles, and we shall see that, while the power system comes near to meeting our requirements for a simple system, the control system is more complex. This is not a surprising discovery when the success of insects in colonizing dry environments with only intermittent supplies of food and water is recalled. Under such circumstances the spiracles open sufficiently for adequate gas exchange to take place (and the insects may be active) while restricting water loss to a minimum.

Many of the results have been gained from extracellular electrical recordings made from the spiracle nerves, from connectives (whole and after splitting) between ganglia and from other apparently associated nerves. They have been carried out under a variety of conditions which include localized gassing and temperature changes, destruction of parts of the peripheral and central nervous systems, and alteration of the bathing medium. An approach of this type may seem like that of a man who attempts to discover the nature of the government of a country by studying the rules which control the air conditioning

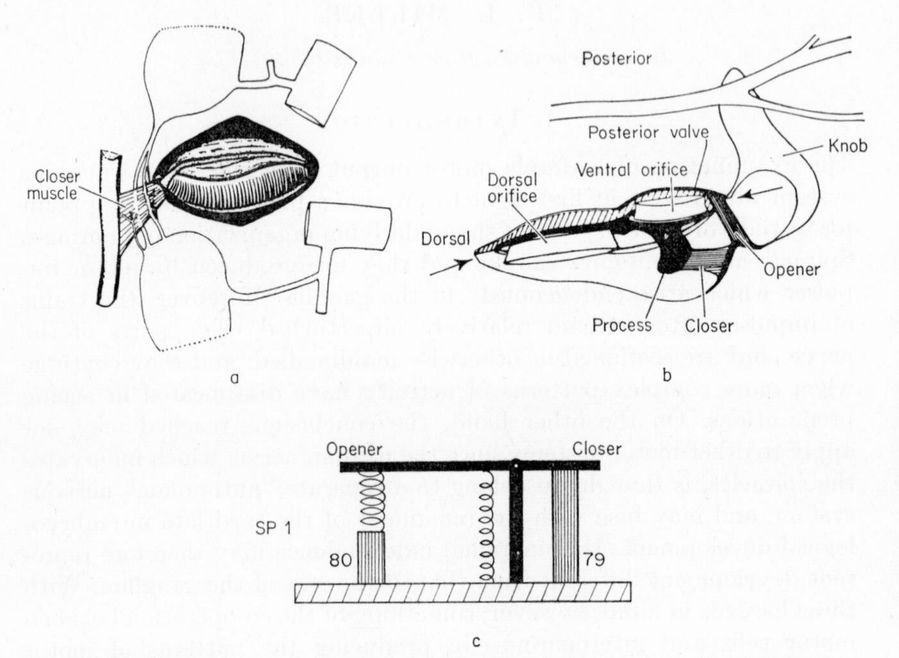

FIG. 1. a, spiracle 2 (mesothoracic) of an adult dragonfly (Anisoptera), viewed from the inside. b, spiracle 1 (prothoracic) of the desert locust. Arrows indicate the axis of rotation of the posterior moveable valve (hatched). c, a model representing the action of the opener and closer muscles of spiracle 1 of the locust.

systems in its debating chambers. However, attempts are now being made to hear the debate in progress, with intracellular electrodes.

The spiracles to be described are the first (prothoracic) in the desert locust and the second (mesothoracic) in several species of adult dragonfly (Anisoptera). Spiracle 1 in the locust has a posterior moveable valve which can be closed by a broad muscle (the closer) and opened by a more slender antagonist (the opener). The opener acts through a cuticular spring which stores energy until the closer relaxes (Miller, 1960b). Spiracle 2 in the dragonfly has one muscle only, the closer. It is opened by a cuticular spring when the closer relaxes (Miller, 1962), (Fig. 1).

II. The Motor Supply to the Power System

In both the dragonfly and the locust the median nerve leaves the mid-dorsal region of the segmental ganglion and then divides, sending two motor axons to the closer muscle on each side. In the absence of ventilation the motor cells of these axons are continually active under normal conditions, and they fire at slightly different frequencies. Records from the spiracle nerve, therefore, show spikes coming in and out of phase at regular intervals. This behaviour, which is here termed "free running", usually occurs at a frequency characteristic for the spiracle; for example, in the locust it is commonly at 17–19/sec per axon in spiracle 1 and at 8–10/sec per axon in spiracle 2 (Fig. 2). Free running

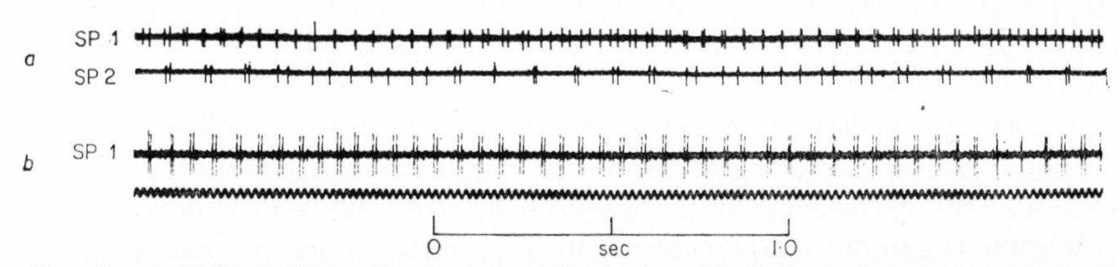

Fig. 2. a, activity in the motor nerves to spiracles 1 and 2 of a locust recorded simultaneously. The metathoracic ganglion has been removed and "free running" continues with all four motor cells firing independently. b, an unusual example of free running in the spiracle 1 nerve, in which the two cells fired nearly synchronously for more than 1 hour.

continues with little change in frequency after the ganglion is isolated by section of the anterior and posterior connectives and of the lateral nerves, although in certain species of dragonfly there may be a significant change following decapitation (Miller, 1964a). In fresh preparations there seems to be no coupling between the firing of the two cells in one ganglion or between those of adjacent ganglia. In older preparations of the locust spiracle 1, however, the cells may drift into phase and then remain firing synchronously for several seconds before they drift out again. In one preparation synchronized firing continued for over an hour.

Free running at a frequency of about 8–10/sec per axon may give rise to alternate periods of fluttering, coinciding with synchronous firing, and of full closure of the valve, coincident with the arrival of spikes out of phase. This is commonly seen in spiracle 2 of the locust after section of the cord between meso- and metathoracic ganglia. Fluttering has been reported in the spiracles of many insects. The synchronization of impulses in the two axons for long periods might be a means of achieving continual fluttering (i.e. minimal opening) without a change in frequency of either nerve cell.

In dragonflies free running may be periodically interrupted by patterns of activity superimposed from ventilation "centres" in the abdominal ganglia. For example, in *Hadrothemis* a short high-frequency burst of spikes, coinciding with the expiratory stroke, is superimposed, but free running still occupies the rest of the cycle. In other dragonflies, *Anax* for example, expiration is accompanied by a brief cessation of activity in the spiracle nerve, while in others again (e.g. *Pantala*) both a high-frequency burst and a subsequent silent period accompany expiration, but still free running fills most of the cycle. When ventilation is weak in resting dragonflies, there are no superimposed patterns, and free running continues without interruption. The superimposed patterns bring about spiracle movements synchronized with ventilation, although in the *Hadrothemis*-type valve movements appear only in the presence of CO_2, which causes the closer muscle to relax partially during the lower frequency of free running. In the connectives of the cord activity has been recorded which is synchronized with the periods of high frequency and with the inhibitions. It seems likely, therefore, that in *Pantala* an excitatory and an inhibitory interneurone run from the abdominal ganglia to the motorneurones of the thoracic spiracles, and that they "drive" these neurones for part of each cycle when ventilation is strong (Miller, 1962).

In the locust, patterns are superimposed for much of the time, and free running seldom appears in intact insects. These patterns are derived from the metathoracic ganglion, and it is not until the connectives between meso- and metathoracic have been severed that continuous free running occurs in the spiracle 1 and 2 nerves. The ventilatory cycle may be divided into three phases: inspiratory, expiratory and a pause. The pause often divides the expiratory phase into two parts (Fig. 3). During inspiration the spiracle 1 motor neurones to the closer are silent, and the valve opens. With expiration a high frequency of impulses (up to 250/sec per axon) accompanies the expiratory stroke, while during the pause a lower frequency (about 40/sec per axon) is maintained. Thus, free running may appear in no part of the cycle, all three types of activity being "driven" from the metathoracic ganglion. Even when ventilation ceases for several minutes in oxygen, the driven pattern continues. In some preparations free running may contribute to the activity during a pause, but this seems to be an unusual occurrence.

During the high-frequency burst the two motor cells of spiracle 1 may fire synchronously or slightly out of phase, but there is no synchroniza-

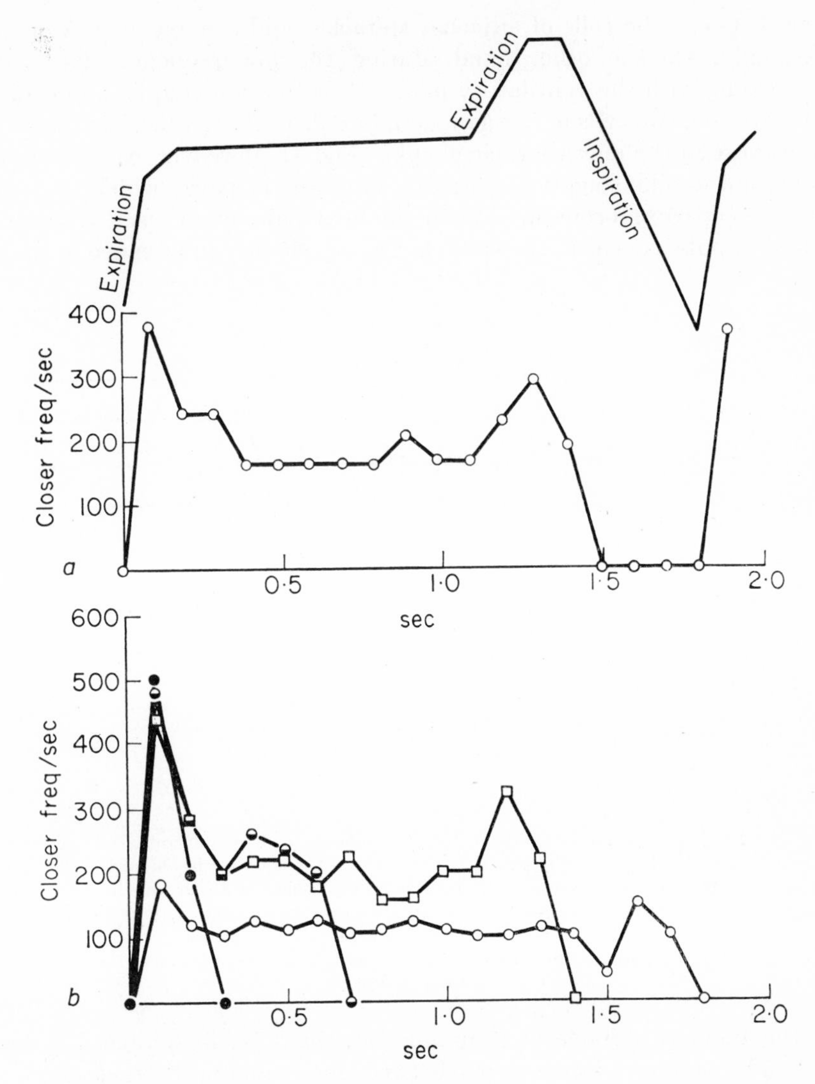

FIG. 3. a, the correlation of abdominal pumping movements (upper line) with changes in the frequency of motor impulses (lower line) in the nerve to the closer of spiracle 1 of a locust. b, the frequency of motor impulses to the closer during one ventilatory cycle in different gas mixtures. ○—○ and □—□, in air. ◒—◒, in 3% CO_2. ●—●, in more than 5% CO_2. With increasing concentration of CO_2, the ventilatory pause disappears, and the activity during expiration is reduced to a very brief high-frequency burst of impulses.

tion between the cells of adjacent spiracles, and a burst may occur in one only. On the other hand, during the low-frequency discharge coinciding with the ventilatory pause, there is often coupling, not only between the two cells in one ganglion, but also between those in the pro- and those in the mesothoracic ganglia (Fig. 4). Moreover, each of these four motor cells sometimes fires its impulses in pairs, which may be 10 msec apart and separated from the next pair by 45 msec. Synchronized impulses occur in the spiracle 2 axons slightly in advance of those

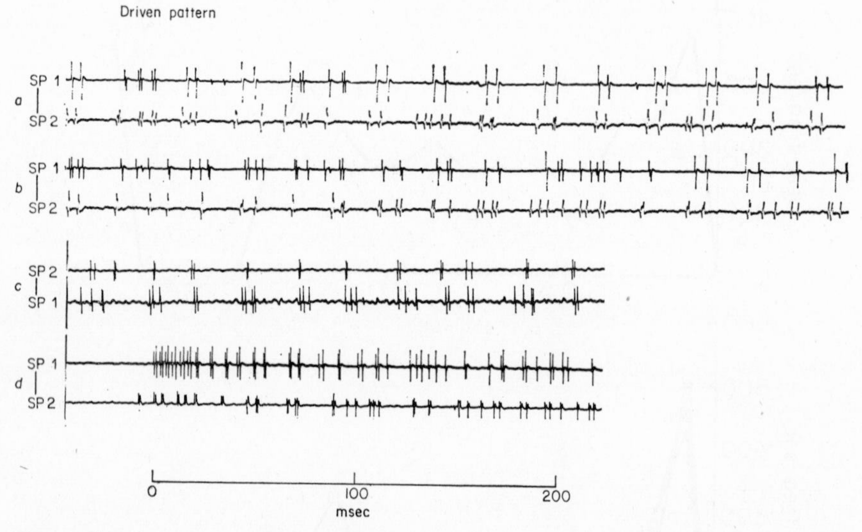

FIG. 4. Simultaneous recordings from the motor nerves to the closer muscles of spiracles 1 and 2 in a locust. a, synchronization of impulses in the two spiracle nerves during a ventilatory pause. Each large impulse represents the fusion of a pair, one in each axon. Impulses in the spiracle 2 nerve usually occur slightly in advance of those in the nerve to spiracle 1. b, as above, but with looser coupling between the two spiracles and between the motor cells of each spiracle. c, as above. d, a high-frequency burst at the beginning of expiration.

in the axons to spiracle 1. Assuming that the delay in spiracle 1 is due to conduction in an interneurone between meso- and prothoracic ganglia and that synaptic delay in each ganglion is similar, a conduction speed for this interneurone of 3–4 m/sec at 25° C can be postulated.

During a ventilatory pause the two cells again fire either synchronously or a few msec apart, and when the latter occurs one cell leads for long periods. For example, the record shown in Fig. 4a has the following sequence for spiracle 1: S S, S AB AB, S S, S S, S AB, S AB, S S, S S, etc. (S = synchronous; AB = cell A firing in advance of cell B; commas indicate grouping). During free running, leadership regularly alternates between the two cells, and this provides one method of distinguishing the two forms of activity. Figure 4 shows that the incidence of synchronous as opposed to the AB type of firing in the motor cells of one spiracle bears no relation to its occurrence in the other. Moreover not all the spikes in one spiracle nerve are matched by spikes in the other, that is the coupling is often loose.

When the connectives between the pro- and mesothoracic ganglia are cooled slowly to 0° C the various driven patterns are blocked in turn. First the high-frequency burst disappears; secondly, the low-frequency driven pattern is gradually replaced by free running; and finally, the inspiratory inhibitions disappear. Free running now continues indefinitely, and all inputs from the metathoracic ganglion are blocked. As the preparation returns to room temperature the patterns reappear in the reverse order.

It seems possible, therefore, that three interneurones run from the metathoracic ganglion and make synaptic contact with the motorneurones of spiracle 1. These are (i) an inhibitory interneurone which is active during inspiration; (ii) an excitatory interneurone which fires during expiration (the lack of synchronized firing between adjacent ganglia suggests that a separate interneurone supplies each); (iii) an excitatory interneurone which is responsible for the maintained pattern of activity during pauses in ventilation. Synchronization of firing in pro- and mesothoracic ganglia under these circumstances suggests that one interneurone supplies both, and possibly, in addition, the more posterior ganglia. This interneurone seems to be absent in dragonflies where free running fills the gaps between successive ventilatory strokes. Since the behaviour of spiracle 1 differs from that of spiracle 2 in flight, separate interneurones (i) and (ii) must supply each spiracle.

III. Control Systems in the Dragonfly

In the dragonfly, control can be exercised only through the closer muscle. Normally free running produces a tetanus of the muscle, and the valve remains shut. Locally applied CO_2 reduces this tension, and the valve may start to flutter, the mechanism probably being similar to that in spiracle 2 of the locust (Hoyle, 1960). If there is now an increase in the frequency of motor impulses arriving at the closer, it will again close, and if a high frequency is maintained the sensitivity to CO_2 will be reduced. The frequency varies with temperature (with a Q_{10} of 2·7) and also with the water balance of the insect. When water is short the frequency is raised, and in consequence there is a reduction in the sensitivity of the spiracle to CO_2. Temperature and water balance changes produce a smooth alteration of the frequency of free running, and they may act directly on the motor cells of the spiracles. In contrast, hypoxia decreases the frequency by interrupting free running so that it becomes irregular; in an undesiccated insect free running may be inhibited altogether in 2% oxygen. Receptors in the head ganglia or in thoracic ganglia can initiate this reaction (Miller 1964a, b), and an interneurone seems to be involved.

IV. Control Systems in the Locust

A. *Control via the Closer*

The frequency of free running in a locust with the metathoracic ganglion removed is decreased by perfusion of the head with CO_2 or hypoxic mixtures. Free running may be inhibited completely under these conditions. In contrast, perfusion of the tracheal system of the prothoracic ganglion produces no effect unless much stronger CO_2 concentrations are used (15–20% in oxygen). Lower concentrations of CO_2 in decapitated insects may *increase* the frequency of free running, and there is sometimes a greater effect on one cell than on its partner, so that the apparent "beat" frequency is increased.

Stimulation of the eyes by light, or of any part of the head mechanically, during treatment with CO_2 permits free running to reappear. It may continue for 15–20 sec after the end of the stimulation before it is again completely inhibited. Such stimuli applied to the head may inhibit the activity of the interneurone which is in turn inhibiting the prothoracic closer motorneurones.

In the intact insect the activity which is driven from the metathoracic ganglion is not affected by perfusions of the head with gases. Their effect is on ventilation and on the activity of the opener muscle of spiracle 1. Control via the closer may therefore be of little significance in the intact ventilating insect, but it is of interest that by removal of ventilation we can uncover in the locust a system which has a functional counterpart in the intact dragonfly.

B. *Control via the Opener*

Earlier investigations revealed the presence of two motor axons running in the transverse nerve to spiracle 1 and supplying the opener muscle. They arise in the prothoracic ganglion (Miller, 1960b). A third axon supplying this muscle has recently been found. It starts in the mesothoracic ganglion, travels anteriorly in the first lateral nerve and then

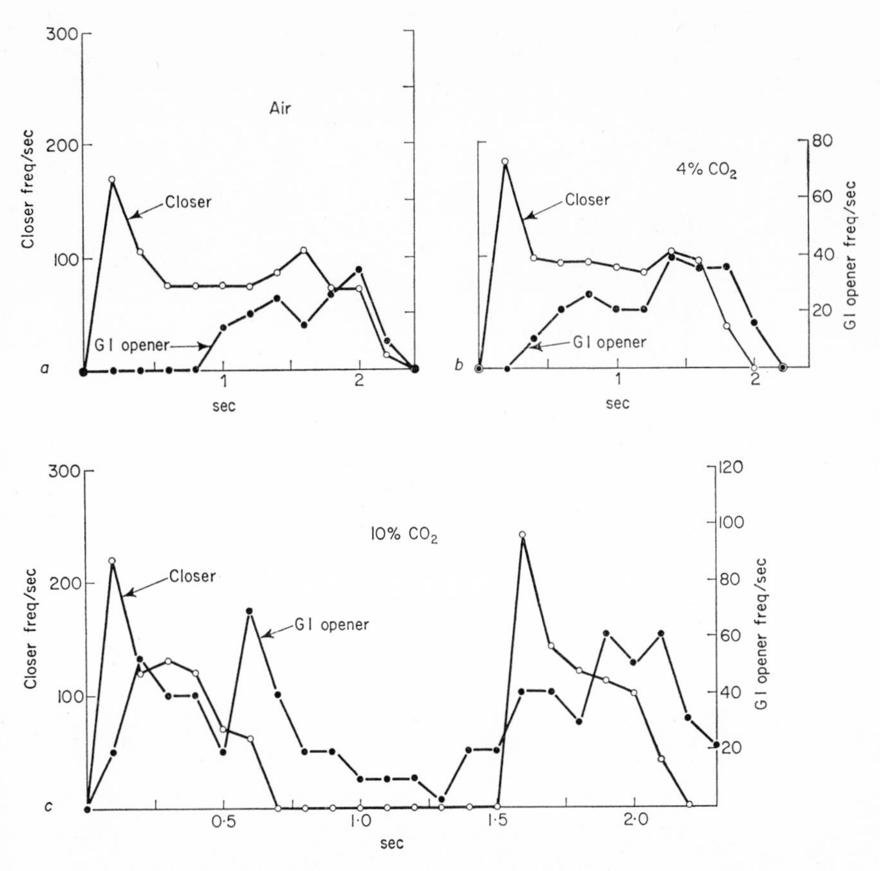

FIG. 5. The frequency of impulses in the motor nerves from the prothoracic ganglion to the opener and closer muscle of spiracle 1 of a locust in different gas mixtures. a, in air. b, in 4% CO_2. c, in 10% CO_2.

joins the transverse nerve (Fig. 5). Impulses in these three axons cause contractions of the opener muscle, so that the valve gapes more widely when the closer relaxes. They may remain silent or fire at low frequencies when ventilation is weak in resting locusts. The prothoracic nerves are stimulated by CO_2 or hypoxia in the head or in the prothoracic ganglion, whereas the mesothoracic nerve is stimulated by CO_2 or hypoxia only in its own ganglion.

When ventilation is moderately strong, the prothoracic nerves to the opener fire during the period of closer activity, that is, during expiration. Opener impulses may start after those in the closer nerve, but they cease simultaneously (Fig. 6a). Nevertheless, the opener contraction outlasts that of the closer, and the spiracle is wide open during the first

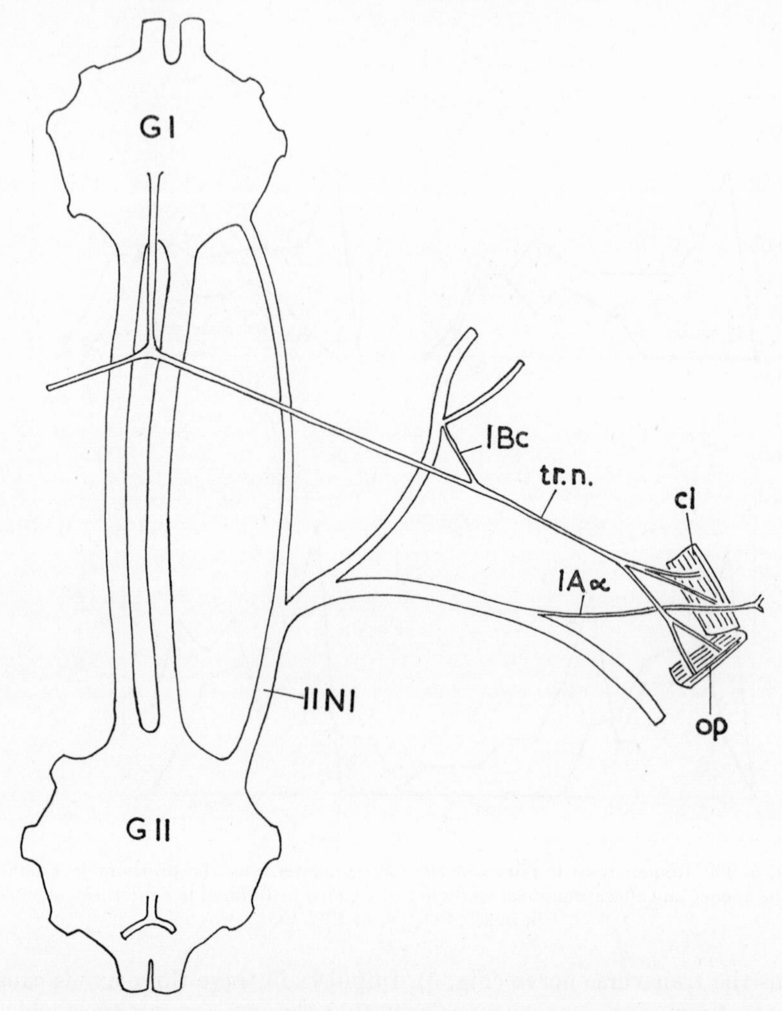

Fig. 6. The innervation of the muscles of spiracle 1 of a locust. The transverse nerve (tr.n.) sends two motor axons from the prothoracic ganglion (GI) to the closer muscle (cl), and two to the opener (op). The mesothoracic ganglion (GII) sends a single motor axon to the opener: it travels in IINI and then joins the transverse nerve via a small branch (IBc). Nerve IAα supplies mechano-receptors in the vicinity of the spiracle.

part of inspiration. This is achieved since the rate of opener relaxation is about five times slower than that of the closer. Moreover, impulses continue to arrive at the opener for about 15 msec after they cease at the closer because the opener axons conduct more slowly (Miller, 1960b). In this way a simultaneous inhibition of opener and closer impulses in the ganglion may be translated into appropriate spiracular action by peripheral delays. During strong ventilation, stimulated by CO_2, the period of activity of the opener neurones may be extended beyond that of the closer neurones, and it sometimes may last throughout the whole cycle (Fig. 6b, c).

The activity of the mesothoracic nerve to the opener occurs out of phase with that of the prothoracic nerves. During moderate ventilation

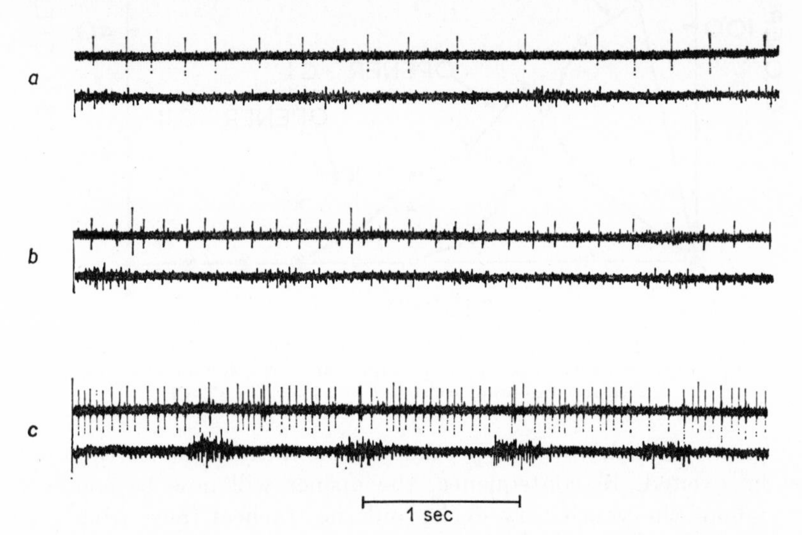

FIG. 7. Activity in the mesothoracic nerve to the opener muscle of spiracle 1 of a locust recorded close to the spiracle after the prothoracic transverse nerve had been sectioned centrally (upper lines). Activity in the expiratory muscle (176) in the third abdominal segment (lower lines). a and b in air. c in 5% CO_2.

the latter fire with expiration, while the mesothoracic nerve fires during inspiration (Fig. 7). In consequence, a contraction may be maintained in the opener by the alternate activity of its pro- and mesothoracic nerves (Fig. 8).

In flight much of the opener activity is derived from the mesothoracic nerve. This means that the opener contracts during inspiration when the valve is open, but it is relaxed during expiration. There is, therefore, no *simultaneous* contraction of opener and closer muscles and no consequent constriction of the ventral orifice of the spiracle (Miller, 1960b). Pterothoracic gases can therefore be driven across the atrium of the closed spiracle during expiration and so into the dorsal orifice, from where they are conducted to the head and thus to the vicinity of the regulators both of ventilation (Miller, 1960a), and of the prothoracic opener neurones. If these gases contain about 5% CO_2, ventilation

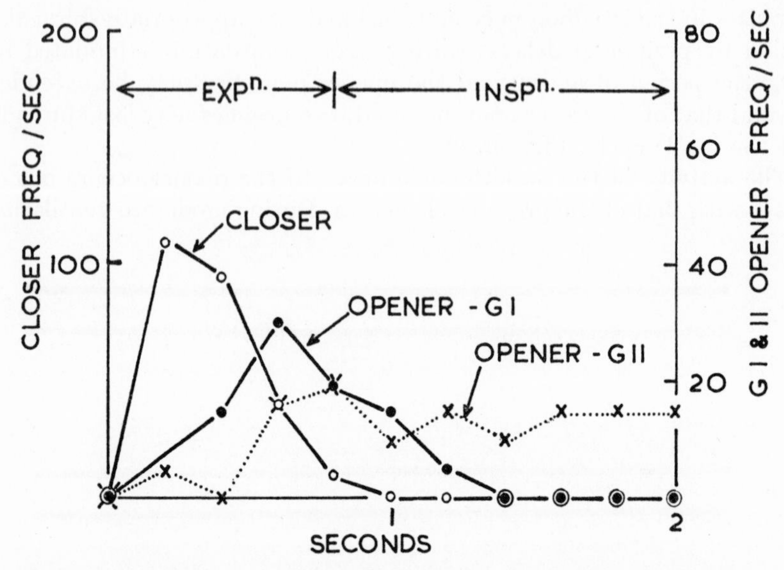

FIG. 8. The frequency of motor impulses during one ventilatory cycle in the nerves to the closer and opener muscles (pro- and mesothoracic contributions shown separately) of spiracle 1 of a locust in 3% CO_2.

will be stimulated, and the prothoracic motorneurones to the opener will be excited. In consequence, the opener will now be contracted throughout the ventilatory cycle, and the tracheal route from pterothorax to head will be constricted during expiration. The further passage of pterothoracic gases anteriorly will be prevented since, as Weis-Fogh (1964) has pointed out, the pterothoracic tracheae form otherwise a remarkably isolated system. The opener activity seems, therefore, to provide a control mechanism operating during flight, which allows the central nervous system to sample pterothoracic gases and to promote appropriate ventilatory responses without itself being flooded by the gases.

In conclusion, we may attempt to summarize in two diagrams some of the findings and the theories concerning the power supply and the control system of spiracle 1 in the locust. In Fig. 9 a possible arrangement of the connections to the closer muscle motorneurones is shown. At this stage it is better not to think of the symbols as representing motorneurones or interneurones so much as component features of the output which seem to be separable. For example, the free running units and other units driven from the metathoracic ganglion are depicted separately, but they may well co-exist in one cell. Again, the three wires between metathoracic and prothoracic ganglia, along which the

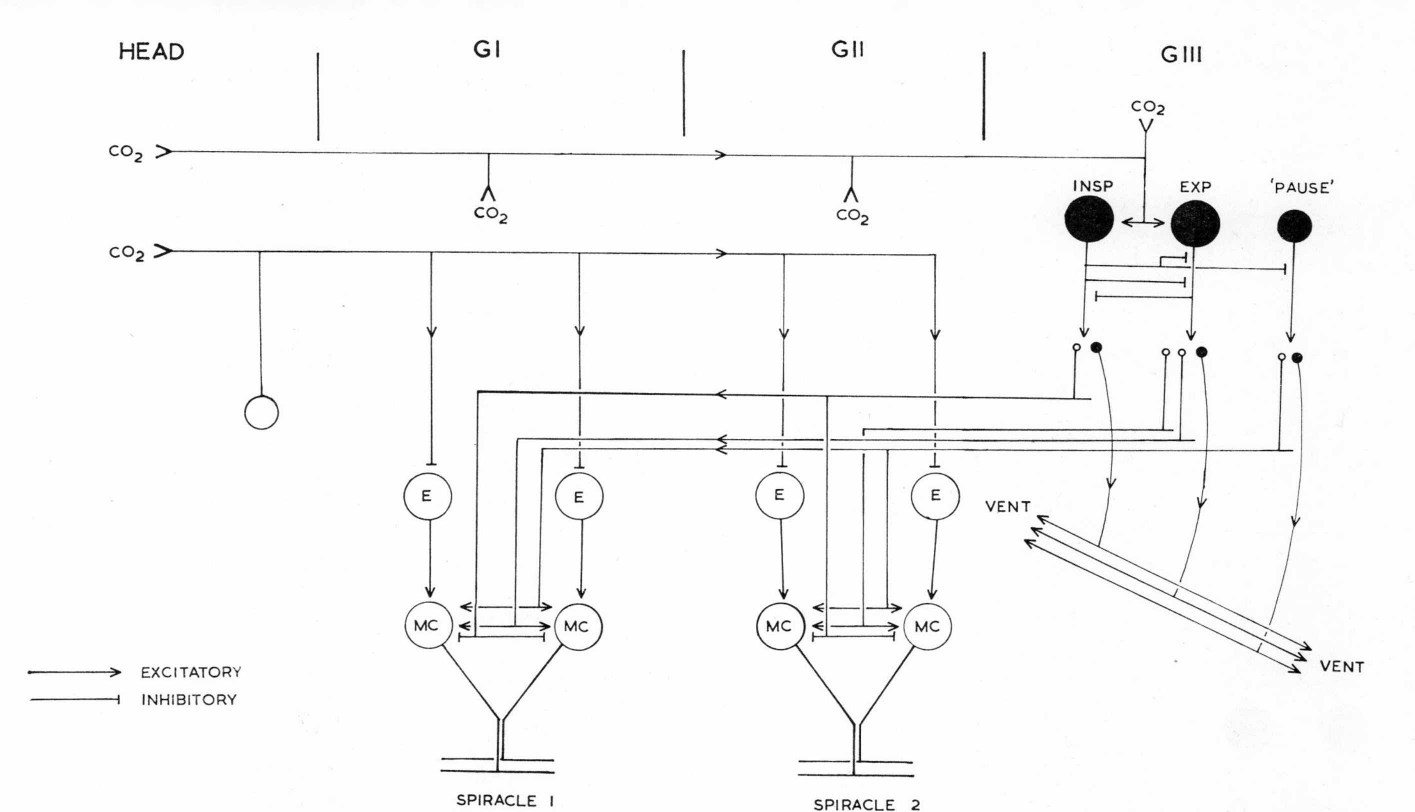

FIG. 9. Hypothetical scheme of the central nervous connections which bring about the motor output to the closer muscles of spiracles 1 and 2 in a locust (the power system). GI, II and III, pro-, meso- and metathoracic ganglia; VENT, ventilatory muscles in abdomen and thorax: INSP and EXP, neurones in the metathoracic ganglion which are responsible for initiating the inspiratory and expiratory pumping strokes; MC, motor cells of the spiracle muscles; E, endogenously active neurone which produces the free running pattern.

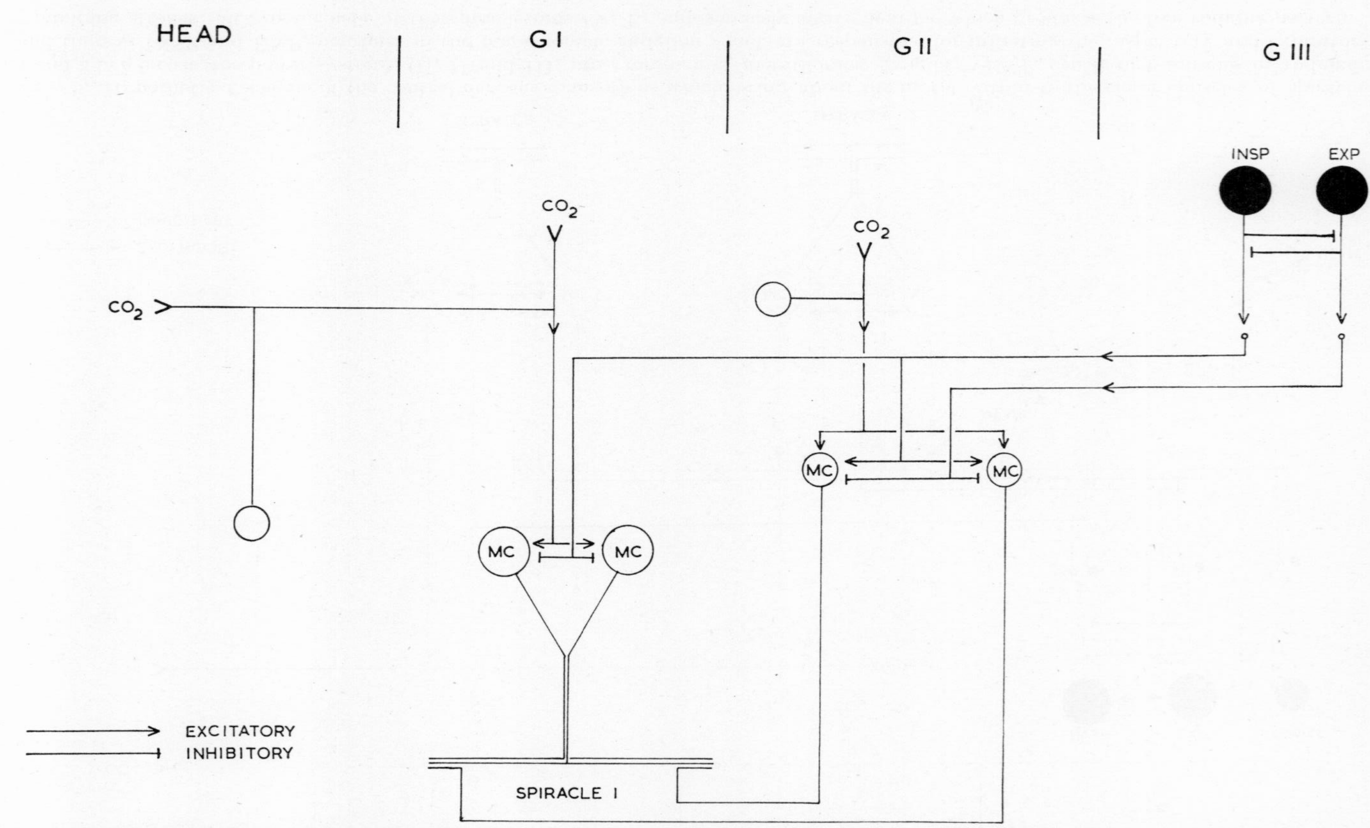

Fig. 10. Hypothetical scheme of the central nervous connections which bring about the motor output to the opener muscle of spiracle 1 of a locust (the control system). Symbols as in Fig. 9.

driven pattern is passed, may represent different aspects of the activity of only two interneurones. The evidence for three is uncertain. Or again, the system may be more complex, involving a greater number of interneurones.

Figure 10 shows a similar portrayal of the evidence for the control of the opener motor cells, and the same caution must be exercised in its interpretation. Such diagrams can be justified only if they summarize a variety of findings with clarity and precision and at the same time suggest further experiments. I believe they meet these requirements.

REFERENCES

Hoyle, G. (1960). The action of carbon dioxide gas on an insect spiracular muscle. *J. Insect Physiol.* **4**, 63–79.

Miller, P. L. (1960a). Respiration in the desert locust. 1. The control of ventilation. *J. exp. Biol.* **37**, 224–236.

Miller, P. L. (1960b). Respiration in the desert locust. II. The control of the spiracles. *J. exp. Biol.* **37**, 237–263.

Miller, P. L. (1962). Spiracle control in adult dragonflies (Odonata). *J. exp. Biol.* **39**, 513–535.

Miller, P. L. (1964a). Factors altering spiracle control in adult dragonflies: water balance. *J. exp. Biol.* **41**, 331–344.

Miller, P. L. (1964b). Factors altering spiracle control in adult dragonflies: hypoxia and temperature. *J. exp. Biol.* **41**, 345–358.

Weis-Fogh, T. (1964). Functional design of the tracheal system of flying insects as compared with the avian lung. *J. exp. Biol.* **41**, 207–228.

The Control of Reflex Responsiveness and the Integration of Behaviour

C. H. FRASER ROWELL

Department of Zoology, Makerere College,
University of East Africa, Uganda

The anatomy of the arthropod nervous system, with its chain of paired segmental ganglia and connectives, encourages an analytic rather than a synthetic view of its functions. While this can cause one to underestimate the importance of the integration of the whole nervous system in normal behaviour, it draws attention to the fact that many units of behaviour (usually and loosely termed "reflexes") apparently involve only limited parts of the CNS. Isolated parts of insects continue to perform many activities which do not differ in essentials from their occurrence in the intact animal. Such activities involve complicated input and output to the nervous system, but the relevant machinery is localized in one or a few ganglia. Similar states exist in all nervous systems, but the morphology of the arthropod version makes it especially suitable for study.

One of the main problems of behaviour is the organization of more or less discrete systems like the insect reflex into a coherent pattern for the animal as a whole. The problem is further complicated in that the "discrete systems" are only functionally distinct, and only then when operational. Their anatomical elements may all be concerned in an indefinite number of other systems as well as the one under consideration. It is fundamentally important to an animal that it should not try to do two things at once, or at least if it does, that they should be compatible. Sensory input is continually arriving on all channels, and there is a real problem regarding which to respond to when and for how long. This is an aspect of what the psychologist thinks of as the problem of "attention", and I hope to give an idea of how I think part of this may be done in insects.

Previous ideas on this subject involve some sort of mutual inhibition between the different operational systems, either between the working parts direct, or between higher nervous representation of these systems. The latter hypothesis, that the "decisions" are taken in the brain, is at least partly true of insects. The results of decerebration are familiar, perhaps the best known study being that of Roeder (1937) on the preying mantis. Removal of the head ganglia increases somatic responsiveness, suggesting an inhibitory influence on the rest of the nervous system. Similarly, Professor Huber (1960) has shown that behaviour can be altered by direct stimulation of the brain, and I have argued (Rowell, 1963) after similar experiments that this is primarily caused by altering the animal's responsiveness to sensory inputs.

However, such a concept of cerebral integration is not adequate to explain the integration of reflexes in insects. Firstly, it merely shifts the site of the mechanism without any suggestion of its nature. Secondly, the decapitated insect is by no means completely disorganized; in insects it is notorious how little difference decapitation often seems to make. Therefore, a large part of the integrative machinery must be intrinsic in the rest of the nervous system. This is true of much smaller units than a decerebrate nerve cord. Related effects from individual cells, where response to a single input is modified by other simultaneous or pre-existing inputs, are already known from insects, as Drs Hughes and Horridge have already mentioned in the symposium. I suspect that mutually inhibitory relations between the different functional organizations are a feature of any multifunctional piece of nervous tissue.

The basic mechanism must be describable in terms of cellular events, but I do not think this is a very useful goal at the moment. A compromise between simplicity and relevance to behavioural events is found in the insect segmental reflex, where receptors, effectors and integrative mechanisms are confined to one segment. All other CNS influences exerted on this system must be through the readily accessible interganglionic connectives.

Such a system can be found in a grooming reflex of the prothoracic segment of acridid grasshoppers (Rowell, 1961); the species I use are *Schistocerca gregaria* and *Acanthacris ruficornis*. The grasshopper makes an oriented movement of the front foot to remove a tactile stimulus from the hair sensilla of the sternal region. All the sensitive areas (except one that can be disregarded for present purposes) are served by the second nerve of the prothoracic ganglion. The motor output to the legs arises from this ganglion only, and cutting the connectives on either side has no effect on the detail of the performance, showing that the integrative mechanism is also in the prothoracic ganglion. By stimulating in a regular fashion a few well-defined sensilla, one can produce an easily repeatable standard input.

With a normal insect there is virtually zero responsiveness; the reflex cannot be elicited. If, on the other hand, all input to the ganglion from the rest of the CNS is prevented by cutting the connectives, and further, direct sensory input reduced by preventing the legs from contacting anything, then responsiveness is almost 100%. This is the simple basis for the experiments: the reflex is inhibited by certain anatomically localized inputs to the ganglion.

To analyse these effects, progressive lesions to the nerve cord were made and the reflex responsiveness tested (Rowell, 1964). Any input from the supraœsophageal ganglion, whether or not it was itself deafferented, resulted in so active and unpredictable a preparation that tests were impossible. The influence of the remaining parts of the CNS was plain, and as an example I should like to consider here only the effects of the posterior ganglia.

Figures are the averages of eight individual performances, each individual being tested twenty times for each lesion.

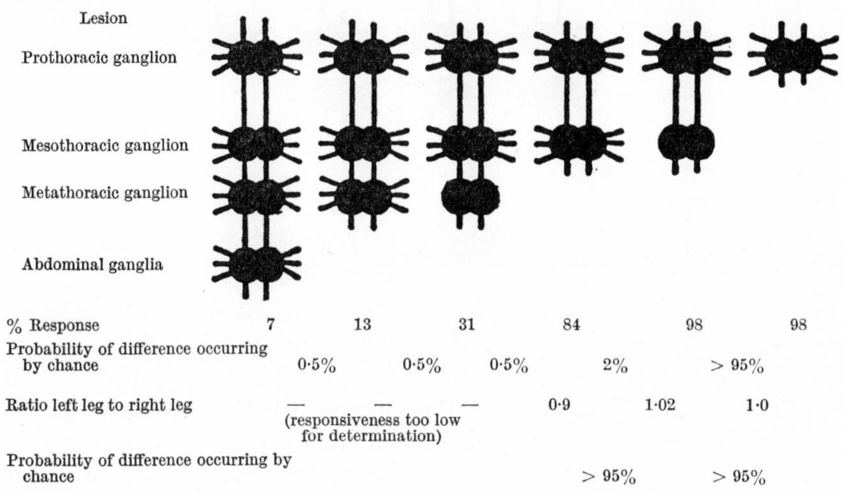

FIG. 1. Responsiveness of the prothoracic reflex after symmetrical lesions to the posterior nerve cord. Head ganglia disconnected. (From Rowell, 1964.)

Figure 1 shows the results of tests on a preparation in which the subœsophageal/prothoracic connectives have been cut, and the only important input to the ganglion is coming from the posterior chain. All CNS behind the neck is intact. Responsiveness is low at 7%. When the CNS is progressively reduced from the posterior end forward, little effect is detectable until all the abdominal chain is removed; responsiveness is thereby doubled (13%). When the metathoracic ganglion is deafferented, responsiveness rises to 31%, and when it is completely disconnected, to 84%. Deafferentation of the sole remaining ganglion, the mesothoracic, causes a slight further rise to 98%, and total disconnection of this ganglion gives no significant advance.

Superficially it would then appear that most of the inhibitory input from the posterior cord is derived from the metathoracic ganglion, and little from the mesothoracic. However, there is reason for thinking that such a deduction is wrong. If one makes asymmetrical lesions to the same area, remembering that each half-ganglion is a largely independent unit, though with good communications with its other half, then a more complex situation is revealed.

Figures are averages of seven individual performances, each individual being tested 20 times for each lesion.

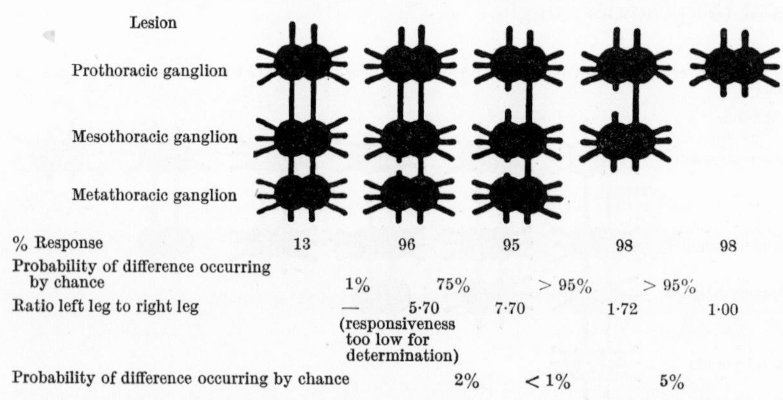

FIG. 2. Responsiveness of the prothoracic reflex after asymmetric lesions to the posterior nerve cord. Head ganglia disconnected. (From Rowell, 1964.)

Figure 2 shows the effects of asymmetrical disconnection of the posterior ganglia. A single unilateral disconnection of the metathoracic ganglion has profound effects. Responsiveness rises straightway to near maximal. However, whereas before the animals on the average used right and left legs equally often, now the leg on the same side as the lesion is used more than 5 times as often as the other. A further asymmetrical lesion has no significant effect upon overall responsiveness, but it does exaggerate even further the unilateral disinhibition (ratio of left leg to right leg now 7·7). Symmetry of response can be restored by cutting the connectives on the intact side. Thus it seems that the mesothoracic ganglion is exerting an important inhibitory influence on the prothoracic ganglion, in spite of the results of the symmetrical lesions, because its asymmetric disconnection leads to a 25% increase in the responsiveness of the operated side. There is also a further paradox: why does a single unilateral lesion (metamesothoracic connective) have more overall disinhibitory effect than a bilateral lesion does at the same level?

A plausible model explaining these effects can be made by postulating negative feedback (mutual inhibition) between the two halves of the prothoracic ganglion. One would expect this feedback to be present to give stability to the system (for example, in its absence there would be no reason why a grooming reflex by one leg should not excite a response from the other, thus leading to oscillation), and there is some independent evidence for its existence too. An example of the working of this model is as follows, considering the mesothoracic ganglion and the prothoracic ganglion only. In the intact state, inhibitory signals pass along both connectives and inhibit primarily their own halves of the prothoracic ganglion. Because of this, the inhibition due to the contralateral half of the prothoracic ganglion is small. If one connective is now cut, that side of the prothoracic is disinhibited. As a consequence, it sends a larger inhibitory signal to the contralateral side, thus decreasing still further its own inhibitory input, and inhibiting more than usual the contralateral side. When both connectives are cut, there is no inhibitory input from outside, but as both halves are disinhibited they exert an increased but equal disinhibition on each other. By this sort of process both these paradoxes can be resolved, and the effects of both meso- and metathoracic lesions accounted for (Rowell, 1964).

This supports Weiant's (1958) postulate of negative feedback between the two halves of the cockroach metathoracic ganglion, which she made to explain rather different results, and suggests that the objections brought to her specific case by Roeder *et al.* (1960) are invalid.

Details aside, it is clear from the above results that lesions of varied sorts, having as their common feature only that they lead to a generalized reduction in activity in the nervous system, result in a progressive disinhibition of the prothoracic grooming reflex. I should like to mention two of the various possible explanations for this. The first is that the grooming reflex is particularly labile (indeed, it was chosen for this study because of this characteristic) and it may be that especial provision is made for suppressing such a reflex. For example, a single interneurone synapsing with a majority of elements in the nervous system and having a specific inhibitory output to the grooming reflex would account for the results. However, specific inhibitory connections, when replicated to the degree required to integrate the entire animal's behaviour, are potentially a wasteful way of using neural elements, and I am reluctant to accept such an hypothesis unless driven to it. This is especially relevant to insects, which have, relative to other types with comparable behaviour such as cephalopods and vertebrates, rather few neural elements (though there is evidence that the small number of cell bodies may be compensated for by unusually profuse branching and numbers of axons).

The second hypothesis runs as follows. Consider a section of nervous system which has many functions, and consider one particular system, such as the reflex described. The reflex is dependent upon the nervous system receiving certain input from the environment. We can call this particular input the "relevant" input—relevant to our investigation, not necessarily to the animal. All other input to the nervous system, in this case the ganglion, is for our purposes "irrelevant". I would suggest that the responsiveness to the standard relevant input is inversely related to the total amount of irrelevant input to the same integrative system. I also suggest that the inhibitory character of the irrelevant input is not primarily influenced by the nature of the input, only by its amount. This is not, of course, to deny that specificity of inhibitory input must occur. It is clear that for such a system to be useful some inputs must carry more weight than others, and this is known to be true. The sensory input from the foot to the prothoracic ganglion does not compare, in terms of signal content, with the input from the metathoracic ganglion, but is if anything more inhibitory in its effect on the prothoracic reflex. Presumably, the function of this particular example is to have the animal avoid using the leg for grooming when it is supporting its weight upon it or feeding with it, etc. Similarly, a learning process involving a reflex of this sort would involve a change in "priorities" of inputs.

There are obvious semantic and other difficulties in trying to define "nature" and "amount" of nervous input. Does one mean total electrical or metabolic power involved, possibly by a DC field effect, or the information content, however that may be measured, or the number of action potentials, or what? This can only be usefully debated after an examination of the nature of the inputs and the way in which they are changed in the lesions described. Thus, for the lesions described one should analyse the input passing along the pro-mesothoracic connective during the experiments.

At the moment, an electrophysiological and anatomical study of this input is being carried out, and the former immediately raises a major technical problem. The solution has some interesting theoretical implications. An electrode placed on the intact pro-mesothoracic connective will record both *up* and *down* signals—action potentials propagated in both directions. We are interested in only one of these signals. The usual way out of this difficulty is to cut the nerve, but in the present case this may have serious effects on the signal which one is trying to obtain, in that the input to the mesothoracic ganglion will now be greatly altered. Simple tests confirm that this is so. If the prothoracic ganglion is progressively isolated while recordings are made from the cut anterior connective, then a progressive reduction in activity will be found until the last nerve is cut and the ganglion is completely isolated. After a few minutes, intense activity commences in most ganglia, an activity remarkable for the number of units firing and the regularity and patterning of their discharge. This continues for up to 25 minutes, after which time the ganglion dies suddenly (possibly from exhaustion of food reserves). This result stresses the importance of maintaining normal connections during investigation. It seems that the ganglion is not designed to be stable in the absence of input, and if this state is produced experimentally, there is a real danger of mistaking the "spontaneous" patterned activity which will result in these abnormal conditions for a normal functional operation of the nervous system.

For my own purposes, this question has been solved by developing an electronic device which allows one to distinguish the up and down traffic in a nerve without cutting it. The analysis has just begun. If the input to the prothoracic ganglion from the intact cord is expressed as a histogram of spike size with frequency, then my hypothesis would demand that the lesions will result only in an overall decrease of frequency of all elements, not in a change of shape of the histogram. A drastic change of shape would suggest that one type of fibre was especially concerned with the inhibitory input, as well as suggesting possible specific inhibitory connections. So far, perhaps because of having to use such a crude measure as spike size, there has been no evidence to support this case, or to suggest that the inhibition is due to one specific fraction of the population of neurones comprising the input.

This does suggest that the inhibition is a function of the total signal arriving as my hypothesis would demand; however, this conclusion is still very tentative.

REFERENCES

Huber, F. (1960). Untersuchungen über die Funktion des Zentralnervensystems und inbesondere des Gehirnes bei der Fortbewegung und der Lauterzeugung der Grillen. *Z. vergl. Physiol.* **44,** 60–132.

Roeder, K. D. (1937). The control of tonus and locomotor activity in the preying mantis, *Mantis religiosa* L. *J. exp. Zool.* **76,** 353–374.

Roeder, K. D., Tozian, L., and Weiant, E. A. (1960). Endogenous nerve activity and behaviour in the mantis and cockroach. *J. Insect Physiol.* **4,** 45–62.

Rowell, C. H. F. (1961). The structure and function of the prothoracic spine of the desert locust, *Schistocerca gregaria* Forskål. *J. exp. Biol.* **38,** 457–469.

Rowell, C. H. F. (1963). A method for implanting chronic stimulating electrodes in the brains of locusts, and some results of stimulation. *J. exp. Biol.* **40,** 271–284.

Rowell, C. H. F. (1964). Central control of an insect segmental reflex. I. Inhibition by different parts of the central nervous system. *J. exp. Biol.* **41,** 559–572.

Weiant, E. A. (1958). Control of spontaneous efferent activity in certain efferent nerve fibres from the metathoracic ganglion of the cockroach. *Proc. 10th Int. Congr. Ent.* **2,** 81–82.

Extracellular Recordings from Single Neurones in the Optic Lobe and Brain of the Locust

G. A. HORRIDGE, J. H. SCHOLES, S. SHAW and J. TUNSTALL

Gatty Marine Laboratory and Dept. of Natural History St Andrews, Scotland

This is a report of the types of units distinguished in recordings from the brain and optic lobes of the migratory locust during exploratory investigations, together with some of the methodological problems encountered. Indium-filled, platinum-plated Pyrex microelectrodes as described by Gesteland (1963) were used. The locusts (*Locusta migratoria*), reared under constant light, were almost always used intact. The head and thorax are placed on a shaped plastic block and rigidly secured there with wax. So treated the animals will live for many days, and individual large units can be held for the whole life of the preparation if required.

The pioneer studies, and almost all later work, in this field have been by Burtt and Catton (1954, 1959, 1960). In their early work upon acuity they recorded from fairly large units which responded to movements of 0·16° of a bright light, with fatigue on repetition, unaffected by dark adaptation, the visual field being the whole eye. These correspond to our C units. Measurements of latency of responses to flashes (Burtt and Catton, 1959) produced responses simultaneously from numerous units. This study revealed the minimum latency in any particular area although it was not possible to say what type of unit was excited. In 1960 Burtt and Catton distinguished between three types of wide field units in the optic medulla region, (a) "on-off" responses with dark discharge inhibited by light and with high sensitivity to movement of a bright stimulus, (b) "on" units with maintained response in light, silent in darkness and no movement response, (c) "on" units without maintained discharge. These last two types come within our class A. Our work has extended these finds to a wider variety of units by the use of smaller electrodes which record from smaller fibres, by the use of stimulus situations which are intended to bring out differences between units, and by a study of interactions and the effects of

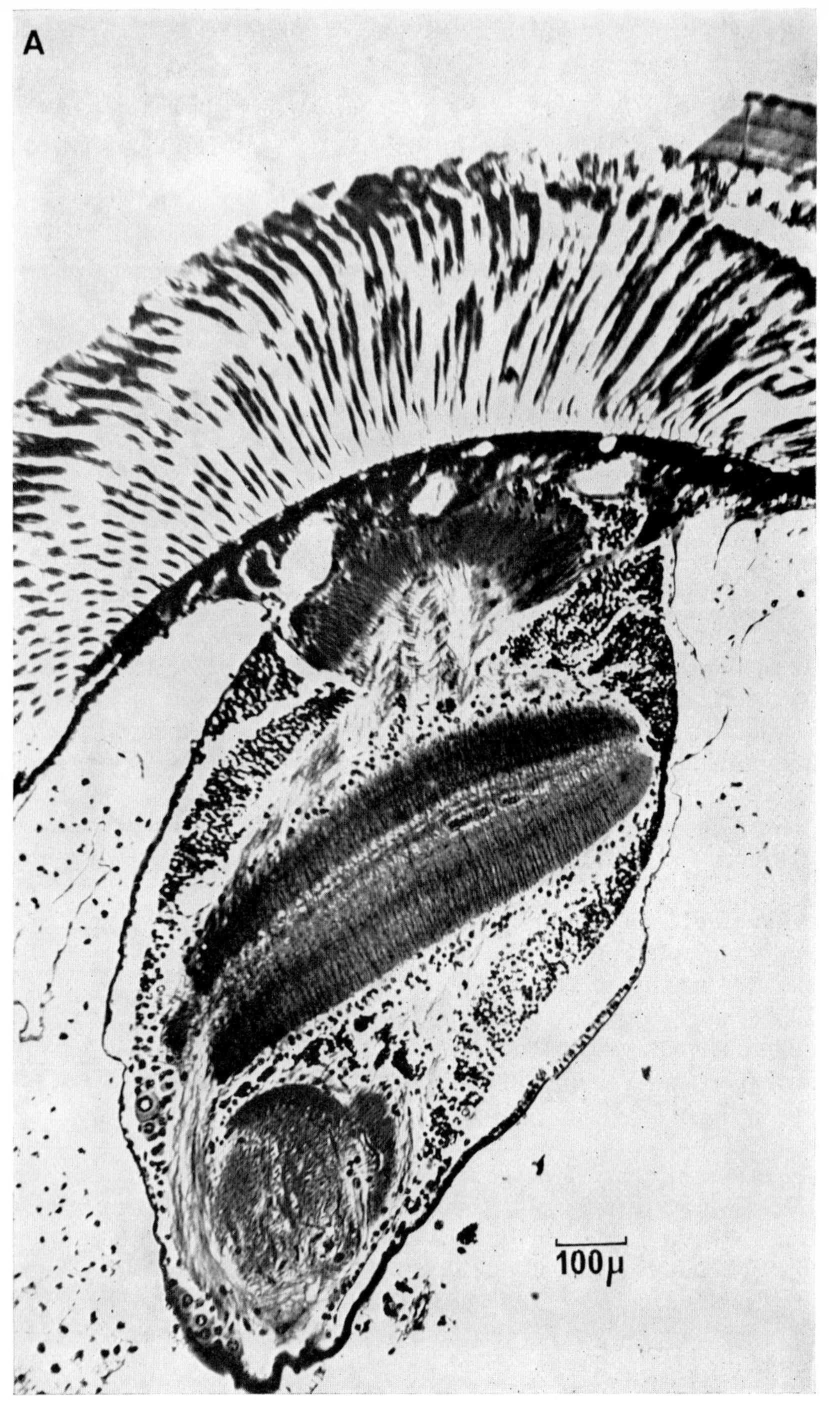

Fig. 1 (A)–(C). Vertical sections through the optic lobe of the locust at right angles to the main axis of the animal. (A) The retina, lamina, medulla and lobula (internal medulla).

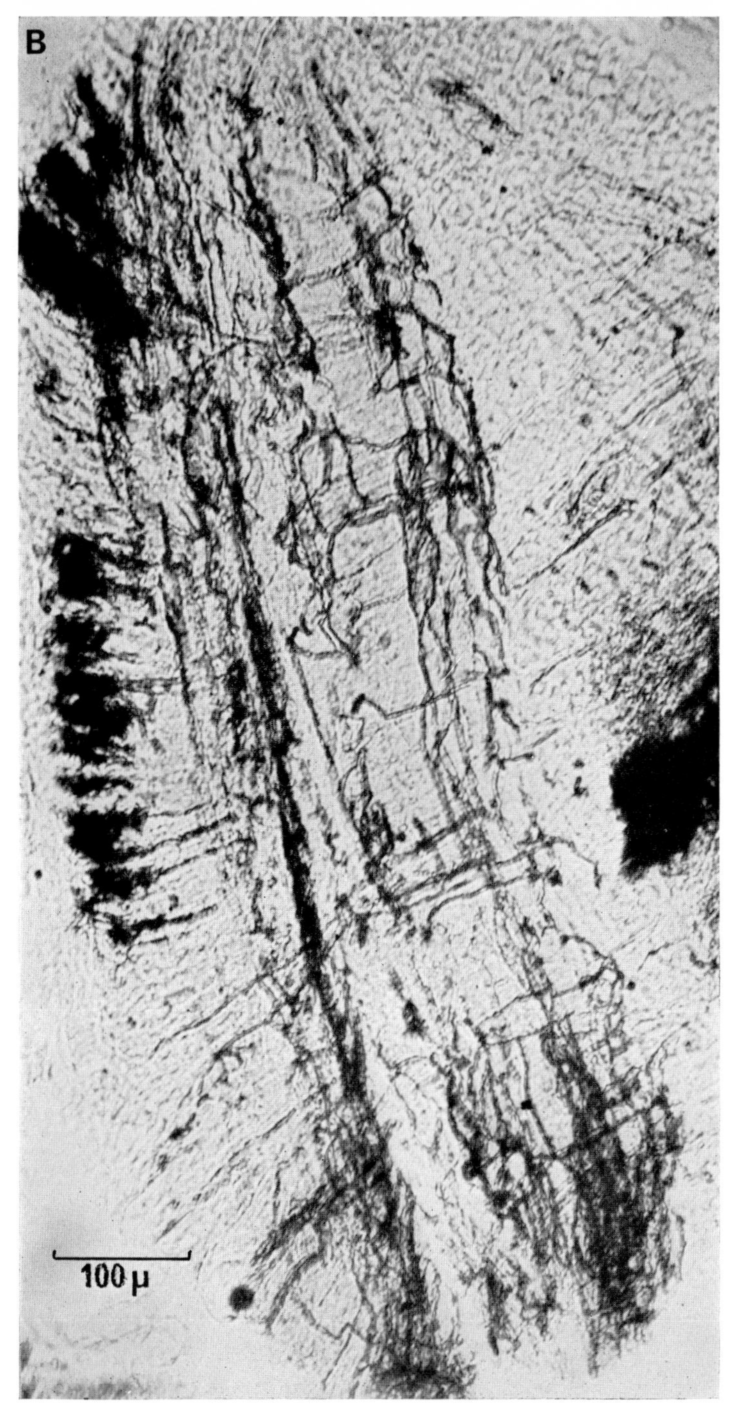

FIG. 1 (B). Golgi preparation of the medulla, showing the layers of horizontal fibres (up and down in this figure) and, at right angles to them, fibres which are radial with respect to the eye.

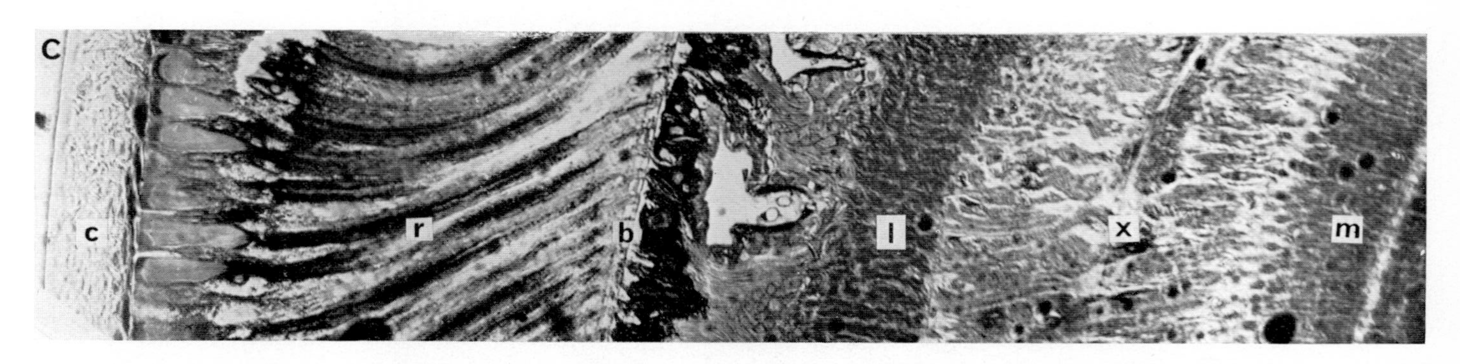

Fig. 1 (C). The line of traverse of an electrode through the cornea (c), retinula cells (r), basement membrane (b), lamina (l), chiasma (x) to the medulla (m).

repetition of the stimulus. The anatomical situation is illustrated in Figs. 1 and 2.

The Retina. At low light intensities retinula cells penetrated with saline-filled microelectrodes yield miniature potentials which seem to

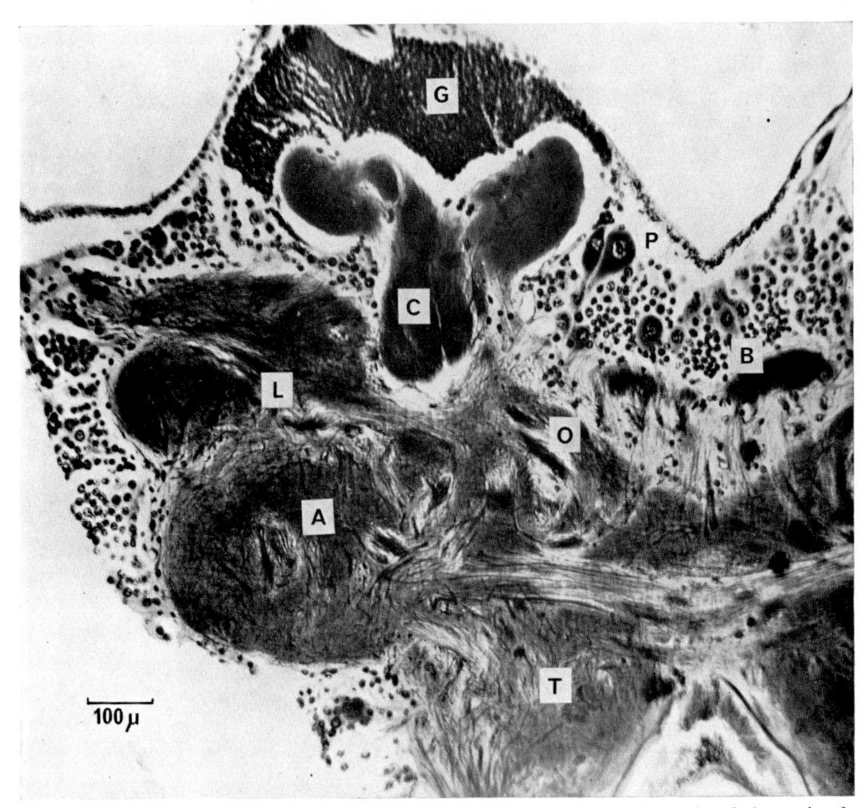

FIG. 2. Section of the brain of the locust transverse to the main axis of the animal, showing some of the regions which were explored. (A) Antennal lobe; (B) Protocerebral bridge; (C) Corpus pedunculatum; (G) Globuli cells of the corpus pedunculatum; (L) Lateral lobe of the protocerebrum; (O) Olfactory-globular tract from antennal lobe to corpus pedunculatum; (P) Pars intercerebralis; (T) Tritocerebrum. The optic lobe is off the picture on the left.

correspond to the arrival of single photons (Scholes, 1964). As the intensity of light is raised the miniature potentials are replaced by a larger graded potential. Spikes, corresponding in frequency to the depolarization of the cell by light, can be recorded from the retinula cells if the electrode tip is near the basement membrane of the retina. Tests have so far revealed no trace of lateral inhibition of the type found in *Limulus*. The difficulty of recording has so far precluded further work.

I. UNITS OF THE OPTIC LOBE, RECORDED EXTRACELLULARLY

The following classification is based upon tests of many hundreds of units, but some, particularly the narrow field units with small spikes, have been found only on particular occasions under favourable recording conditions. Since some units are easily found and others with great difficulty, it is impossible to give realistic figures of the relative abundances.

Class A: Simple total monocular fields, responses of low sensitivity to net luminosity, invariant of factors like contrast, state of adaptation, size of stimulus source or pattern. Responses to repeated presentations are stable, so long as the repetitions are not repeated more often than one per second. Similar types, with monocular or more commonly binocular fields, are common in the brain.

AB: "On-off" transient response, ratios variable. Weak maintained discharge. Sometimes reverberatory (with an oscillatory response to a maintained stimulus). Low sensitivity.

AC: Pure "on", transient response, very low sensitivity, maintained discharge weak or absent. Sometimes reverberatory.

AD: Sustained net luminosity response, slow adaptation, marked rebound. The maintained discharge frequency indicates the level of illumination, and is weak or absent in the dark. Moderate sensitivity.

AE: Sustained net luminosity response; a variant of AD that is transiently inhibited, or accelerates only slowly at "on".

AF: Sustained net dimming response. Mirrors AD.

Class B: Monocular fields limited in extent or different in different areas. The response may be complex or vary with dark adaptation, stimulus position, or stimulus figural qualities. There is a high sensitivity, and responses to repeated presentations are stable. This class has not so far been found in the brain, except in the optic commissures.

BC: Small receptive fields down to 7 degrees, pure "on" transient response of relatively low sensitivity, no maintained discharge. There is sometimes a widely spread weak peripheral inhibition that does not override the centre "on" response (Fig. 3).

BD: Narrow field (15–25 degrees) net dimming response, of high sensitivity without lateral interaction, with slow adaptation and strong rebound. On histological grounds these could be some of the ganglion cells of the lamina. Subtypes as follows:

BDE: No changes in properties on dark adaptation of the eye.
BDF: Acquire post-flash depression in the dark (Fig. 4).

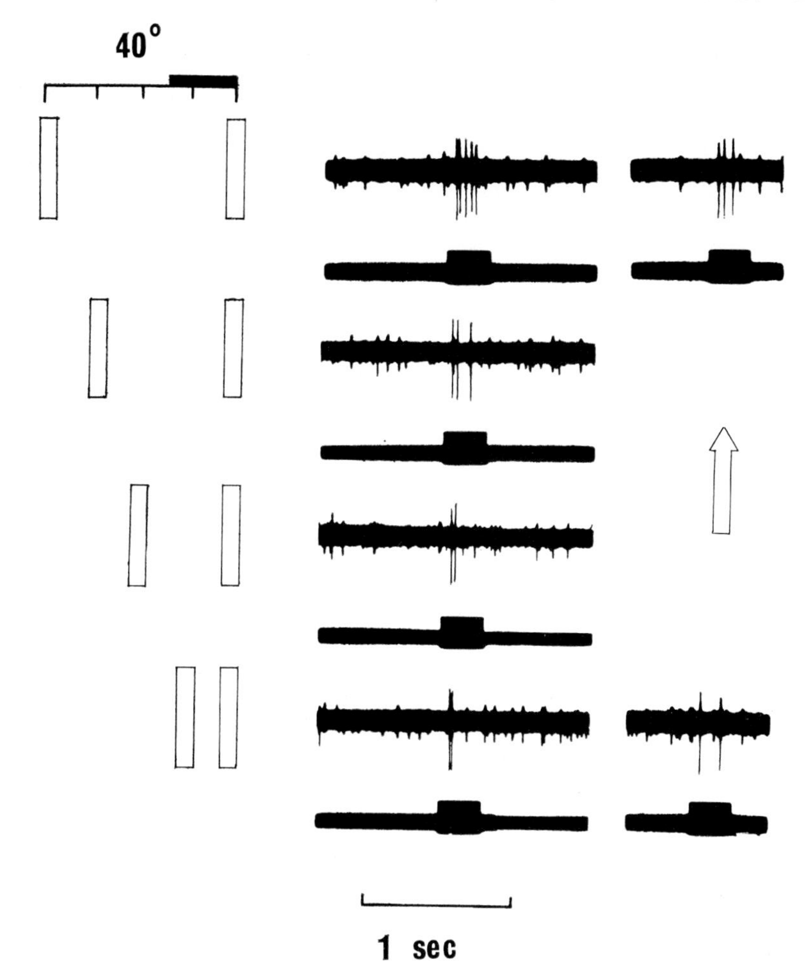

FIG. 3. Marginal example of lateral inhibition. The normal "on" response of a unit to a light flash from the centre of its receptive field is progressively reduced as a second source is brought into the field. On the left of the figure, the successive positions of the light sources are indicated, and the two recordings on the right are taken from a sequence performed in the reverse direction, as a necessary control against habituation. Type BC.

BDG: Response weighted by contrast, and sometimes showing erasability by large changes in background illumination (Fig. 5).

BE: Wide (60 degree) graded field, net dimming response with a complex rebound. Subtypes as follows:

BEF: Pure "on" response in the periphery of the field after dark adaptation.

BEG: Post flash depression in the dark.
BEH: Neither BEF nor BEG.
BEI: Both BEF and BEG.

BF: Twenty degree fields, inhibitory centre, excitatory surround, optimally sensitive to centrifugal movement, maintained background discharge.

BG: Sensitive to the direction of movement, by increase and decrease of maintained discharge with movement in opposite directions (Fig. 6).

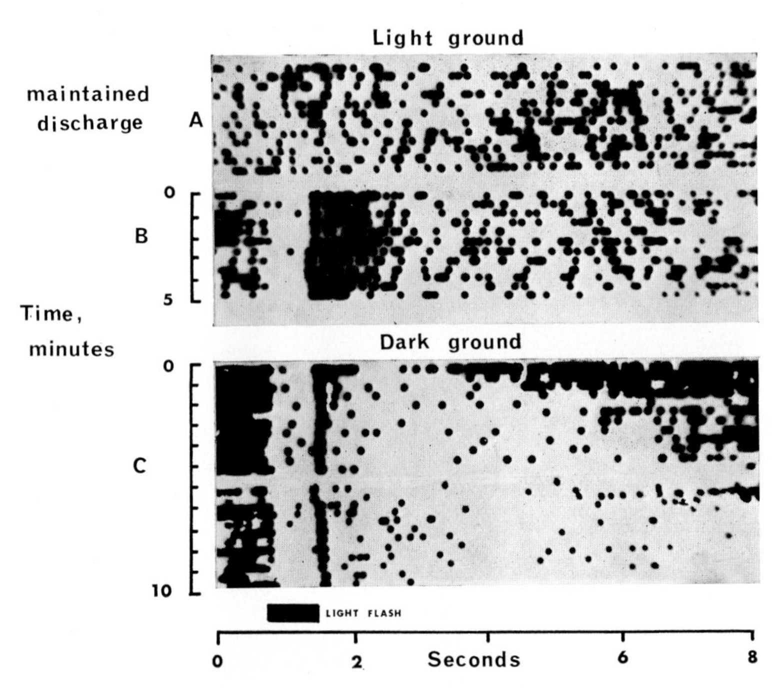

FIG. 4. Changes with dark adaptation. This is a pure "off" unit with a small response field of about 15°. (A) Maintained discharge when presented with a uniform light grey background. (B) The response to a light flash every 30 sec on the grey background is a rebound at "off", with little habituation, and a slow return to a moderate background discharge rate of about 5/sec. (C) As in (B) but all illumination except the flash is now turned off. The response following the flash is now a very short burst and, as dark adaptation progresses, the succeeding depression becomes longer until it occupies 10–15 sec. The background discharge, as seen before the flash, is now much higher than before. Changes of this nature are likely to occur with many of the types of units listed, and characterization in only one state may be considered inadequate once the above phenomenon has been demonstrated, although locusts normally enjoy bright sunlight. Type BEG.

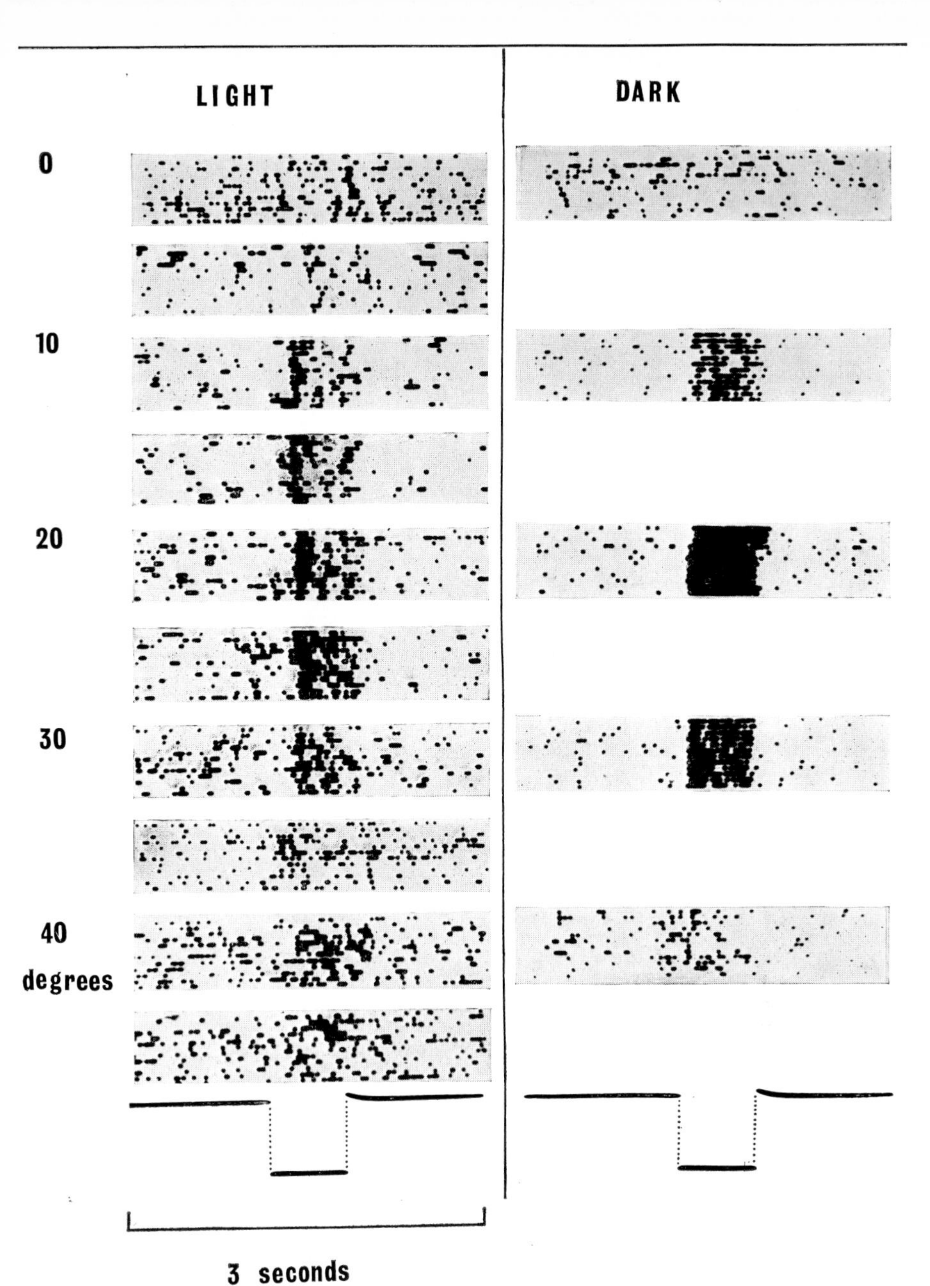

FIG. 5. The form of the receptive field of a sustained pure "off" unit. A green light source (DM 160), normally biased on, is momentarily interrupted by an electrical pulse, as indicated by the monitor trace at the bottom of each column of recordings. In the recordings, dots represent the occurrence of spikes, and each block is composed of a dozen slow sweeps of the oscilloscope beam, which triggers the stimulus. In successive blocks in each column, the light source is moved by 5-degree steps to new positions along the horizontal diameter of the receptive field, so that the density of dots represents its responsive profile. For the left-hand column the light source was positioned against a dimly illuminated grey background. In the right-hand column where all the background illumination has been turned off many minutes previously, the responsiveness is considerably augmented in the centre of the field. This effect is not entirely due to adaptation, because the new average intensity in the field now reduces the background discharge. This effect was not apparent with this unit in experiments involving two small light sources. Type BEH.

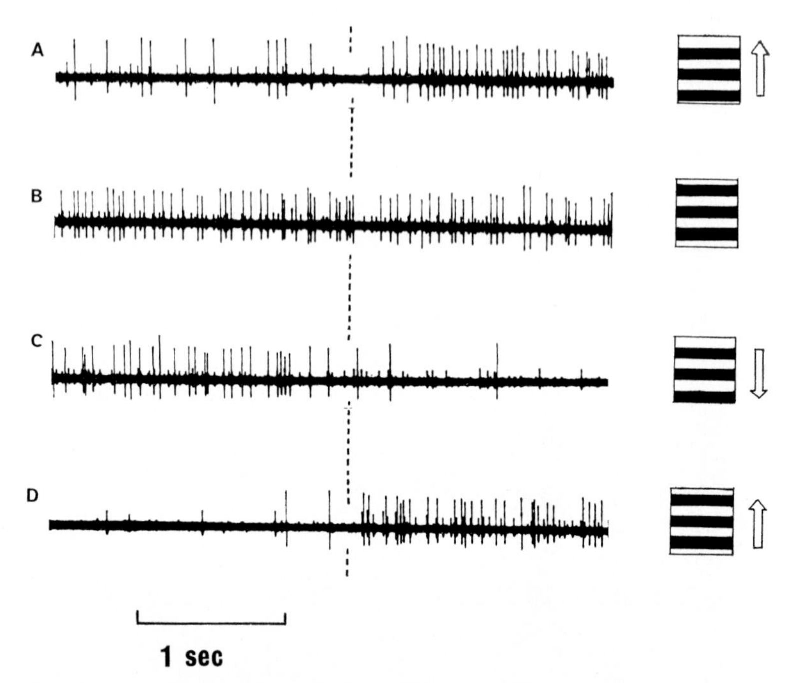

FIG. 6. One-way movement detector. The unit, with a response field of 50 degrees diameter, "looks" through a window at black and white stripes on a drum. In the absence of a changing stimulus situation it has an irregular low-frequency background discharge that is little influenced by the level of illumination, although the size of the window is arranged to give no change in net illumination when the stripes move. (A) The drum is stationary until, at the dotted line, it is moved at constant velocity in an upward direction. (B) The response remains essentially unchanged after a few seconds of movement, until the drum is stopped at the dotted line in the record. There is an inertia in the responsiveness, since the rate of discharge remains greater than background for some seconds after movement ceases. Similar inertia is not apparent at the onset of movement, so that the system shows comparable response to a detector of long time constant and high amplitude non-linearity. (C) The inertial discharge is entirely arrested by movement of the drum in the downward direction. (D) The direction is reversed at the dotted line. The unit does not respond to movement along an inappropriate axis in its field, and apparently confuses several parameters of the effective stimulus, such as pattern, contrast, speed and extent of movement, and the extent to which the movement conforms to the ideal axis. Type BG.

Class C: Total monocular or binocular fields, sometimes converging with auditory and tactile modes, varied sensitivity to illumination changes, sometimes enhanced sensitivity to movement of a contrasting object. Transient responses, adaptation curves sometimes enhanced by inhibition, responses to successive presentations unstable and fatigue away. The rate of habituation depends on the response magnitude, and is independent of the stimulus form, but specific to a visual locus. There is a maintained background discharge, often spontaneously variable. This group are commonly found also in the protocerebrum.

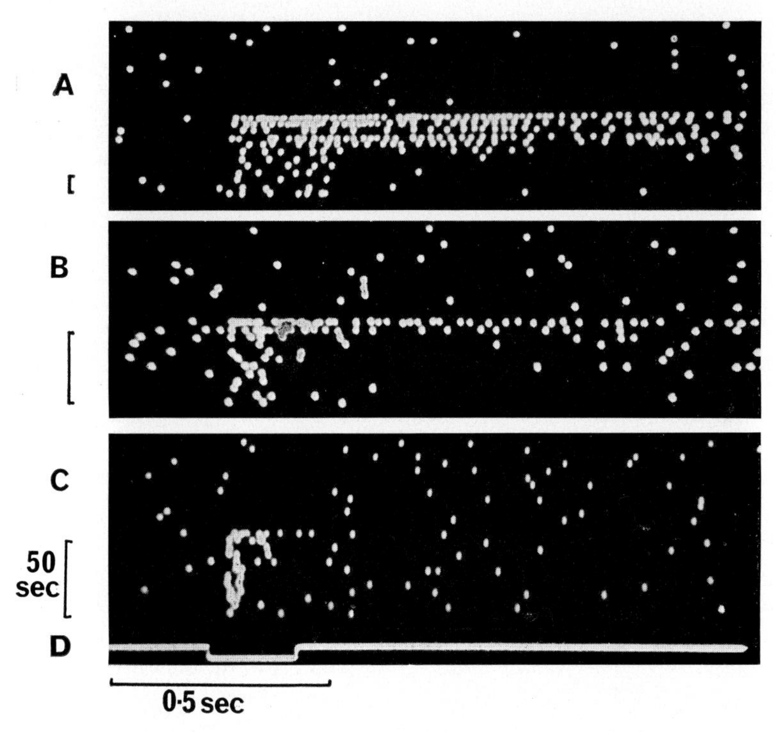

FIG. 7. Habituation in an audio-visual unit. The first part of each record shows the dark discharge in complete silence. Each dot shows the occurrence of a spike. (A) Responses to an ipselateral sound pulse of 10 kc/s repeated every 5 sec, as shown in (D), following a period of many minutes of silence. (B) Responses as in (A), but with a contralateral sound pulse. The response is less and habituates more quickly. (C) Ipselateral responses as in (A), but following a silent period of two minutes, after a contralateral adaptation, showing that sound on one side adapts both sides to repetitions of this frequency. This is the unit with light response illustrated in Fig. 14. Type CE.

CD: Responses of low sensitivity to general illumination steps, contrast not a factor. Movement sensitivity low. Habituation memory short (Fig. 7).

CE: Intermediate sensitivity to movement which causes brightness changes. Typical audio-visual units, not responding to movement *per se*.

CF: Gated against changes of light intensity; highly sensitive to movement, never convergent with other modalities. The form of response is influenced by dark adaptation and by source size. The recovery from habituation to light can extend up to fifteen minutes.

In some audio-visual members of class CD and CE, full recovery from habituation to sound can take many hours.

Class D: Pure auditory, responding as ventral cord auditory units. In the locust there is a greater diversity of units (Horridge, 1961, Fig. 2) than described by Suga and Katsuki (1961) for *Gampsocleis*.

Class M: These are units, mainly found in the protocerebrum, with an intermittent series of bursts of impulses up to 150/sec, not modifiable by any of the stimuli which we tried.

A. *Sensitivity to Other Modalities*

Most of the work has aimed at the visual mechanisms and the interaction between visual and auditory inputs. These types of stimuli have been tested on most units. Chemical stimuli have not been used at all. Tactile and proprioceptive stimuli have been tried on a substantial fraction of units. In general, units of classes A and B do not have inputs from mechanoreceptors at low strengths of stimuli such as gentle touch or blowing. Units of class C commonly have inputs from mechanoreceptors, wind on the head being especially effective. The anatomical areas of inputs from mechanoreceptors have not been examined systematically, but the general result is that the non-specific units of class C are also typically non-specific with reference to mechanoreceptor inputs. There is a class of units, not included here, which respond more specifically to a variety of mechanoreceptors only.

B. *Localization of the Electrode Tip*

The simplest procedure is to enter the eye surface from a known direction and push the electrode through a known distance as measured by a micrometer advance. This allows the repeated reinvestigation of a given area, but in our hands has not proved more accurate than $\pm$ 50μ at a depth of 1 mm from the surface. This is hardly adequate for localization to within each of the neuropile areas of the optic lobe (see Fig. 1). With careful choice of the line of advance from a dorsal point within the opened head, the same method can be useful for identification of the larger areas and tracts in the brain.

A more accurate histological localization has been attempted by examination of sections. About 5V is applied to the electrode, thereby forming a burnt lesion of a few microns in extent at the electrode tip. The scar stains preferentially with basic dyes such as toluidine blue.

The method of histological localization has four disadvantages: (*a*) Accuracy is limited to about 50μ, which will identify the layer or tract, but the method of staining will not identify the neurone type. It is unlikely that this limitation could be overcome by staining with a silver method which would reveal the lesion and the axon type simultaneously. (*b*) From Cajal and Golgi preparations it is clear that most of the larger units, from which one might expect to record, run between different areas, so that localizing them somewhere on the way is not likely to produce a picture of localized functions. Furthermore, the large units which might be identified individually with some hope of success are the very ones which ramify widely in the regions where they occur, as illustrated in Fig. 8. (*c*) An electrode is destroyed by passing the necessarily large current through it, and therefore to localize a unit a preparation and an electrode must be sacrificed. There is, however, a tendency to investigate first one unit then another while recording conditions are good, so that only the last of a number of units can be localized, by which time the interesting units have been lost. (*d*) Even if physiological units in neuropile were identified with histological types in a few cases, it is difficult at present to see the value of this hard-won information. There are about 30 types of unit in the optic medulla alone and in such a neuropile area almost any pattern of interaction may be possible via other branches of the localized neurone. The difficulty is evident from the histological pictures of Cajal and Sánchez (1915), Fig. 8. Moreover, many of the axons in the tracts between areas have not had the branches of their two ends identified together, and information of synaptic relations is almost totally lacking in terms of histological pathways of excitation. There is, therefore, a possibility of analysing the tracts and chiasmata first, and to treat the neuropile areas as black boxes for the time being.

C. *Records from Different Areas*

All the classes of optic units listed above can be recorded in the region of the optic medulla. Many, especially wide-angle visual units, multimodality units and novelty units, are typical of the deeper regions of the optic lobe, the protocerebrum and the optic tracts. Notwithstanding the difficulty of localizing the electrode tip it can be said that the following types of response are characteristic of certain areas (cf. Figs. 1 and 2).

1. *The optic lamina.* Although no units were certainly identified here, many of the small field units of class B are probably of the lamina and first optic chiasma.

FIG. 8 (A–C). Neurones of the optic medulla, from which most of the types of visual units can be recorded, illustrating the dendrites spreading to different distances and the layered strata of horizontal fibres. (A) Neurones (G) with widely spreading dendrites and a radially directed axon. Large horizontal (tangential) fibres (H) have branches which spread throughout the neuropile mass. These neurones could hardly be narrow-field units.

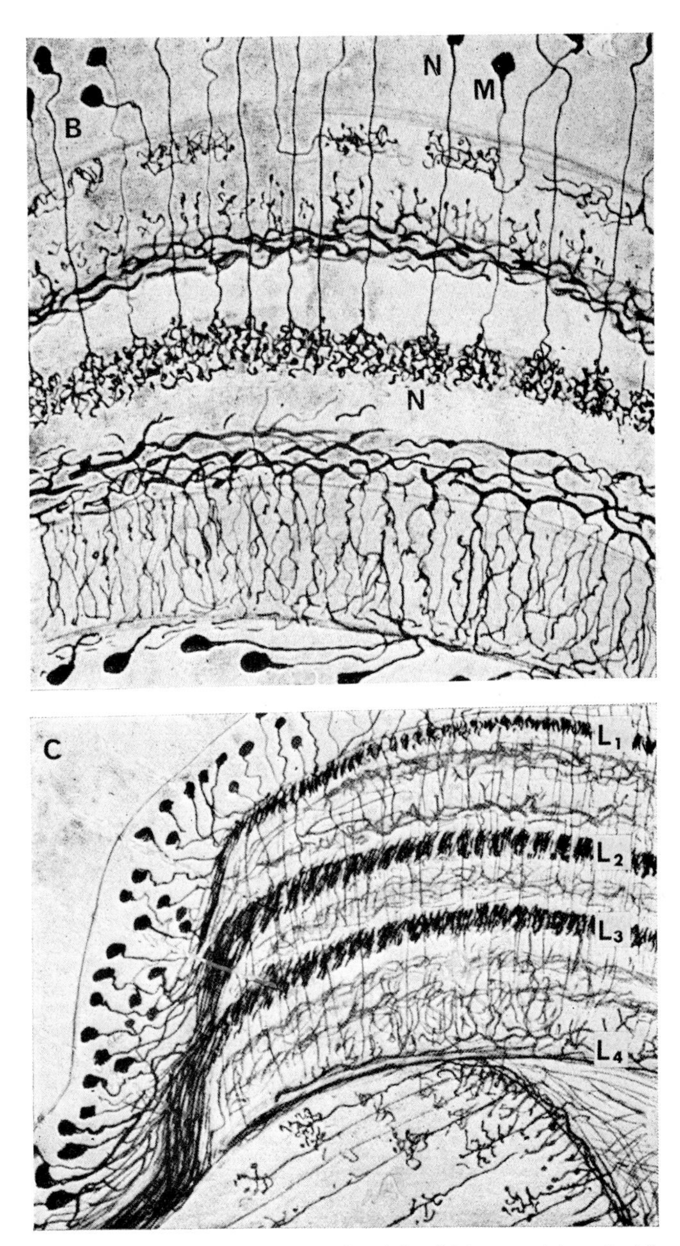

FIG. 8 (B). Small neurones with a restricted dendritic spread (N and M) forming a repeated array across the neuropile mass. These are not likely to be wide-field units. (A) and (B) are from dipteran optic lobes. (C) The layers of horizontal fibres of the medulla of the dragonfly. These fibres form tracts to the brain and to the contralateral optic lobe. Modified from Cajal and Sánchez (1915).

2. *The optic medulla.* Most of the types of visual and audio-visual units can be recorded in the medulla. The histological form and large size of the fibres of the medulla suggest that the large wide field and multimodality units run in the horizontal strata which are parallel to the eye surface (see Fig. 8C).

3. *The lobula* (*internal medulla*). In the locust this region is rather small and not readily distinguishable from the optic tracts. Except that the units with the smallest fields occur only towards the periphery of the optic lobe there is no marked correlation between type of unit and depth in the optic lobe. Presumably this is because most of the axons which yield responses are in optic tracts which run along the length of the optic lobe.

4. *The corpora pedunculata.* These areas showed numerous very small spikes in response to typical visual stimuli, especially to gross changes in light intensity as recorded by Burtt and Catton (1959). There is a considerable projection of optic tracts into the calyces. We did not see slow waves persisting through a period of illumination. The intrinsic globuli neurones of the corpora pedunculata are too small to give responses by our methods. A traverse made by the electrode in the stalks region (α and β lobes) does not yield a clear distinction between the corpora pedunculata and the surrounding neuropile, because probes will often record similar activity for a depth of 0·2 to 0·3 mm whilst the stalk of the c.p. does not exceed more than approx. 0·1 mm. Some probes into this region occasionally record antennal responses and units excited by mechanical stimulation of the labral hairs, probably via the olfactory-globular tract.

5. *The central body.* Probes in this region record the same activity as in the corpora pedunculata, with a larger incidence of responses of deuterocerebral origin. However, the localization of these units must remain in some doubt, and could easily be associated with the olfactory-globular tract or the trito-protocerebral tract, both of which pass close to the central body. A number of large spontaneous M units, firing repeatedly and unmodifiably at 50 to 150/sec, have been recorded here and from the protocerebral bridge, and have been observed for more than an hour with no apparent change in their behaviour.

6. *The protocerebral bridge.* The responses are similar to those of the central body. There are ocellar units which have marked "on" and "off" responses, not readily adapted out, similar to those recorded with hook electrodes from the ocellar nerve itself. As can be seen in silver preparations, the large lateral ocellar fibres pass by this area. Frontal probes in this region often pick up antennal units and a number of multimodal and spontaneous units.

7. *The olfactory-globular tract.* The responses are in general the same as those from the antennal lobe. The majority of units have little or no spontaneous firing rate, but give a rapid, quickly-adapting response when the antennae are mechanically stimulated. Units were recorded which would respond to movements of only the ipselateral antenna and occasionally only to specific areas upon that organ. Responses from the mouthparts were seen rarely. Bimodal units responding to visual and antennal stimulation rapidly habituate, suggesting that we are dealing with an "alarm response".

Optic commissures. Entry is made posterior to the lateral ocellar nerve where the brain and the optic lobe meet. There are numerous large visual units, mainly of types A and C, but almost invariably with binocular field responses. One particular slowly adapting type with a background discharge was excited by a large increase in light intensity falling upon the ipselateral eye, but inhibited by similar stimulation of the other eye. Simultaneous continuous illumination of both sides yields a series of bursts of spikes with gaps in between the bursts. Such units were recorded on only three occasions among many hundreds of other units of the "on-off" variety or the pure "on" type. Contrary to the expectation that they would be confined to the outer layers of the eye, we found occasional small field units of the BG type responding with rapid adaptation to unidirectional movement by modulation of a resting discharge and insensitive to general light intensity changes. Audio-visual units are found occasionally. Of the visual units, spikes in many of the largest "on-off" units could be paired with spikes in the ventral nerve cord, but one-way movement detectors could not be found in the ventral cord, although the optomotor responses of flying insects suggest that they should be there.

Localization of units reveals the following general features. Almost all units in the protocerebrum are responsive to visual stimulation, either solely or less commonly in combination with some other sensory modality, often antennal. The various neuropile areas do not yield separate types of responses, presumably because the records are taken only from large, widely-ramifying fibres. If brain regions are functionally specialized, as almost any theory of the mechanisms in the brain must suppose, then the neurones with localized or functionally specific properties are too small to be recorded with our electrodes.

D. *Ocellar Units*

Ocellar units are found in the brain but not with certainty in the optic lobes. Those of the brain lie around the region of the ocellar nerves and are presumably the central ends of the peripheral (second-order) fibres. "On" and "off" responses of these fibres were as reported by Hoyle (1954) except (following Burtt and Catton, 1959) no units with a maintained dark discharge have been found. Hoyle recorded from whole ocellar nerve, and we possibly missed some of his units. Responses in the protocerebral lobes were of the contralateral ocellus. One new type of unit in the lateral protocerebral lobe shows an interaction of the two lateral ocelli; the "off" response to a contralateral light is abolished by a light stimulus which is switched on at that moment to the ipselateral ocellus. These experiments were done with the compound eyes painted over. The close proximity of the lateral ocellus to the compound eye and the great sensitivity of the latter makes it difficult to separate the responses of the two receptors when both are being tested with the electrode in the optic lobe.

II. Physiological Characters of Units

A. *Habituation*

Visual and mechanoreceptor units always habituate to some extent on repetition of the stimulus, but some auditory units respond with more impulses to clicks at certain intervals, as illustrated in Fig. 9. There is a wide range in habituation characteristics, as measured by the rate of the decline in the response with stimuli repeated at different intervals, and in the recovery curve during a period of rest. Some recover in a few seconds, others require up to six hours. This range demonstrates that most of the habituation effects are of central origin and do not depend on sensory adaptation. Where arousal or hit-and-miss responses have been demonstrated in one or two units, habituation cannot represent a loss of excitation but is more like the closing of a gate which can be opened again. The effect may be widespread, though it has proved difficult to demonstrate. The novelty units appear to work in a consistent way, namely by local habituation to movement without influence on other regions in the field of response of the same unit.

B. *Variety of Responses*

Most units, especially those of the brain, are disappointing in their responses. Whereas a richness in the response to pattern and frequency of presentation of the stimulus might have been expected, one finds mainly a sensitivity to movement, and to level of illumination (or net

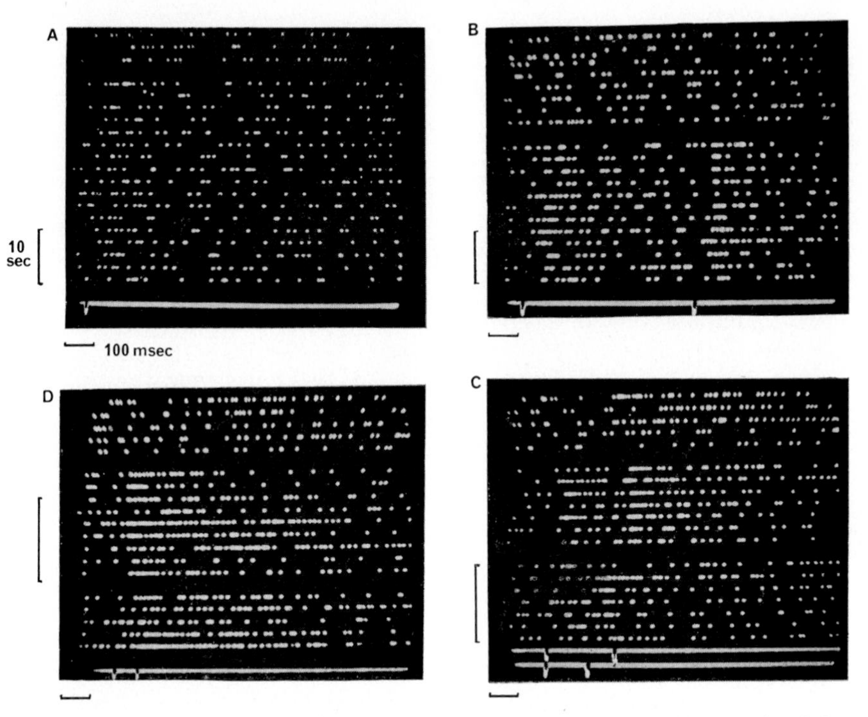

FIG. 9. Facilitation of the response to repeated clicks in a pure auditory unit of the optic medulla. Each dot represents a nerve impulse; the first few lines of each record, as far as the omitted line, show the background discharge. (A) Responses to clicks every 2 sec (one click per sweep) shows little habituation or augmentation to successive clicks. (B), (C) and (D) Responses to identical clicks, except that two clicks occur at each sweep, at the following intervals: (B) 0·55 sec, (C) 0·22 and 0·14 sec, (D) 0·08 sec. Note the persistence of the responses in (D). Light stimuli had no effect on these responses. This type of behaviour was rarely encountered and was not typical of audio-visual units, which characteristically habituate to any repeated stimulus.

luminosity), with wide fields, low sensitivity and typical habituation on repetition. The limitation is almost certainly one of technique. The animal is strapped down; the stimuli, although in the physiological range, are meaningless in the context of the animal's ability to make behavioural responses, and not presented on the background of stimulation which characterizes the animal's natural environment. It is not surprising that almost all recordings can be interpreted as alarm responses dulled by repetition.

Despite this general comment, the most interesting units were the one-way movement detectors, the small-field units with excitatory centre and inhibitory surround, and those with inhibitory centres and excitatory surrounds. The last two are types of movement receptor, which respond to receding and approaching contrast patterns respectively, because they respond best to movement of a bright object from their inhibitory to their excitatory area.

C. *Interaction of Inputs*

Units which respond to several different types of stimuli where different types of sense organs are involved show, in general, little interaction between the several inputs. For example, audio-visual units can commonly be adapted to one type of stimulus without marked effect upon the sensitivity to the other. This can be true even when the two stimuli interact when applied together or in close succession. There are many possible explanations of this phenomenon. The simplest is that the different sensory inputs act on widely separated dendrites of the unit from which the recordings are made.

The same relations are found where wide field visual units are habituated in one small fraction of their field and then tested in another fraction. They behave as if they have ramifying dendrites between which there is little interaction. Novelty units are an example which are typically sensitive to movement and locally habituate rapidly, with long recovery times.

An example of two interacting units recorded simultaneously is shown in Fig. 10.

D. *Auditory Units*

The stimuli were derived from an oscillator which actuated a deaf-aid ear-plug, producing a noise of relatively low intensity, below the threshold of response by hairs. Most units responded at "on", some with short-lived, others with long-lived bursts to clicks. There was no evidence of pitch discrimination. In the case of the large audio-visual units, the receptor was proved to be the tympanic organ by a demonstration that the response disappears the instant the tympanic nerves are cut. Large audio-visual units are readily recorded in the optic lobes, but are rare in the brain. Pure auditory units are rare in all areas and have been found in the brain only in the lateral protocerebrum. Of these a few responded to ipselateral but not to contralateral sound pulses (Figs. 11A and B). They are probably the ventral cord auditory units ascending to the optic lobes. In view of the finding that local stimulation of the corpora pedunculata initiates chirping in crickets and that such chirping can be an alternation between males (Huber, 1962), we expected to find auditory responses in the corpora pedunculata but

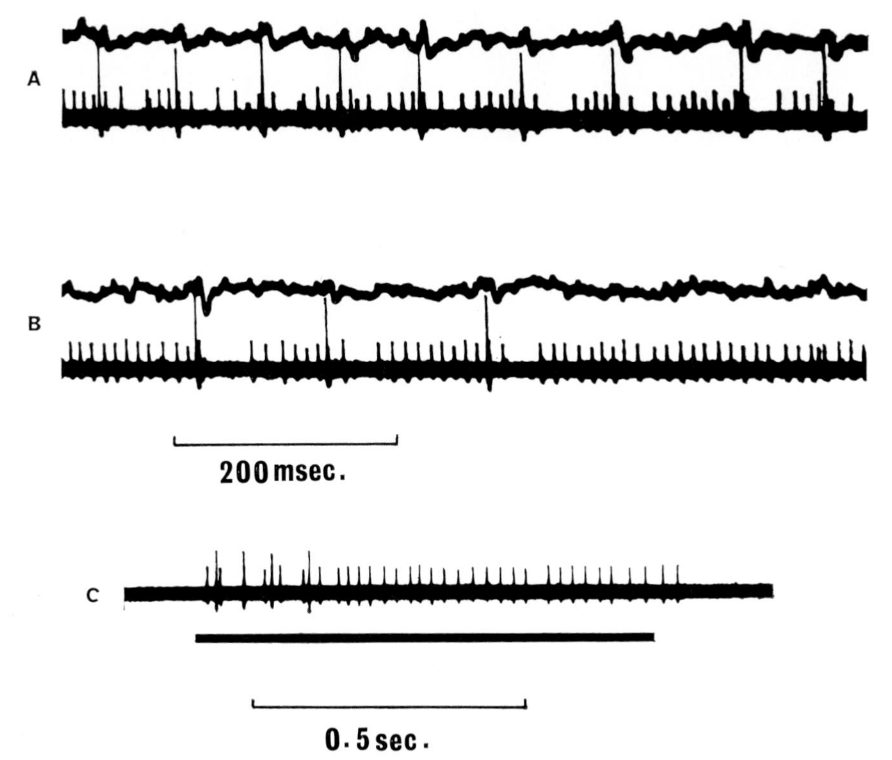

FIG. 10. Interaction between neurones. Two units are recorded simultaneously by the same electrode. Both are pure "on" units, although the one with the larger spike adapts rapidly while the other adapts slowly. Recordings are taken (A) early, and (B) late, during an intense light exposure, showing that following a spike in the large unit there is invariably an inhibitory pause in the discharge of the tonic unit. The upper line in each record is a smoothed version at higher gain, showing the lack of a slow wave of comparable time course to the inhibition. (C) The consequence of interaction during an exposure to light. The tonic unit is allowed to reach its maximum frequency only when the phasic unit no longer fires. Types AC and AE.

failed. Two pure auditory units in the optic lobe had atypical sound responses, one giving a small burst for the duration of the sound pulse, adapting completely after two or three repetitions with a recovery time of about 10 min, and the other giving a strong prolonged burst to a short tone, appearing not to adapt at all. Occasionally an auditory unit with background discharge was slightly inhibited by the "on" of a sound pulse (Fig. 11C), but most responded with a burst of impulses at "on" and responded best to rustling or scraping noises as distinct from pure tones.

Audio-visual units, or possibly the same unit in different animals, differ widely in rate of habituation and recovery from habituation;

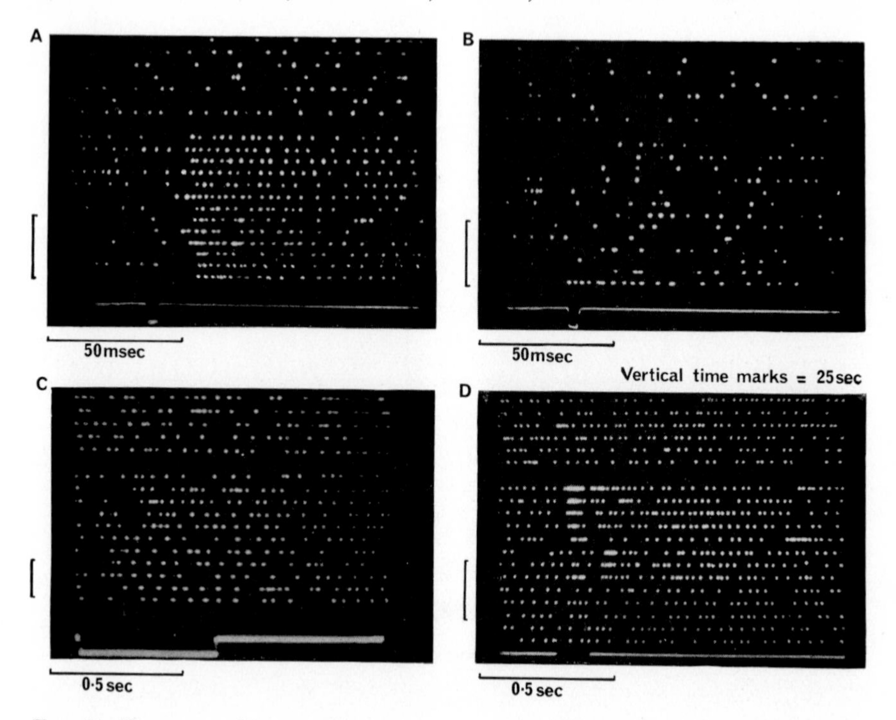

FIG. 11. Responses of pure auditory units. Each dot represents a nerve impulse. The records read like lines on a page; the first few lines of each record show the background discharge in the absence of a stimulus. (A) and (B) Non-habituating response to ipselateral sound as in (A), but no response to contralateral sound as in (B). The sound stimulus in each case was a 5 msec sound pulse of 10 kc/sec delivered from a deaf-aid earphone at 20 cm. The animal evidently has a good directional mechanism. (C) A unit which was perceptibly inhibited at the "on" of a sound stimulus (1·25 kc/sec note of duration 0·5 sec). The response would not be noticeable without the use of the repeated scanning method. (D) Long recovery time from habituation. The sound stimulus was a pulse of 10 kc/sec. After complete habituation a rest period of at least 10 min was necessary to obtain any response, and a period of several hours of silence to obtain a maximal response.

compare Figs. 7 and 12. Sometimes they show hit-and-miss responses which suggest a haphazard gating (Fig. 13A). They always respond to visual stimulation of any part of either eye, and are sensitive to light intensity changes, usually as "on" units, but are not sensitive to movement *per se*. During recording one gains a strong impression that there is a very small population, possibly 2 or 3, widely ramifying audio-visual units and that they lie tangentially in the medulla, where a few characteristic units are always picked up at the same depth of about 1 mm from the cornea. Visually they act as typical wide field units with local habituation to a flash stimulus in one part of the visual field while full sensitivity is retained in other parts of the field, being there fore novelty units for brightness change.

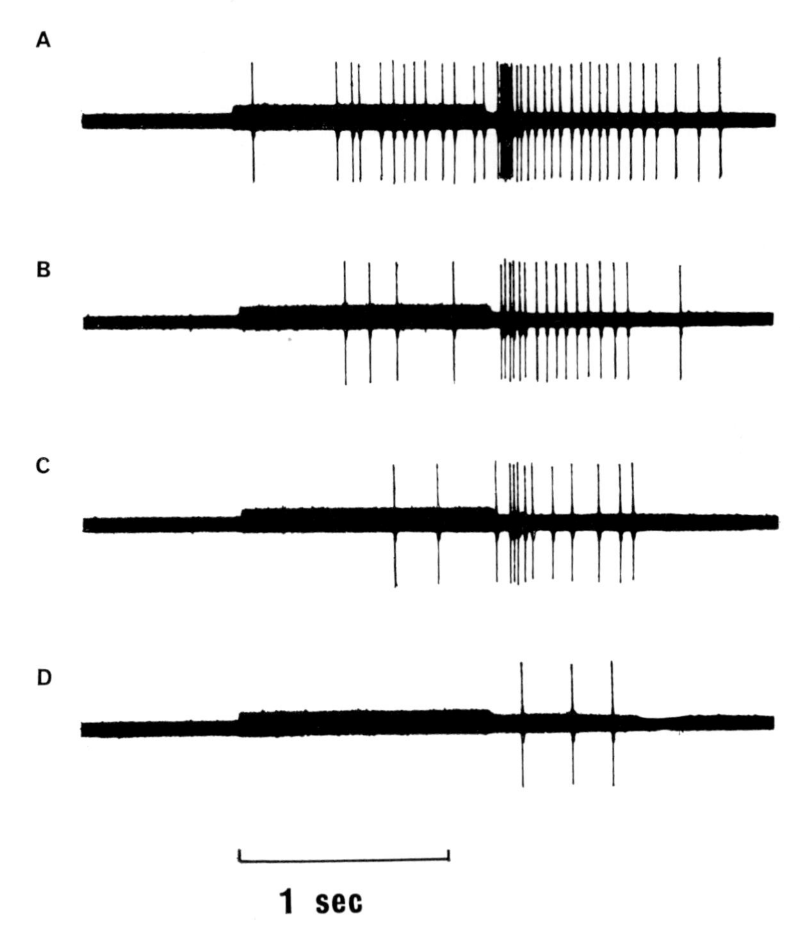

FIG. 12. Habituation in a novelty unit. The four records show responses to successive presentations of identical light flashes which were presented every 5 sec. This unit was rather atypical in its regularity of adaptation; usually responses were less predictable. The long latency at "on", together with the pause after the initial spike in the first line, suggest that the "on" response has been suppressed by inhibition, as in Fig. 10C. Type CD.

E. *Interaction between Light and Sound Stimuli*

Most multimodal units were tested to see whether regularly repeated stimulation via one modality influenced responsiveness to another. Sound pulses were placed in varying temporal relation to light stimuli and the responses examined for differences from the responses to one type of stimulus alone.

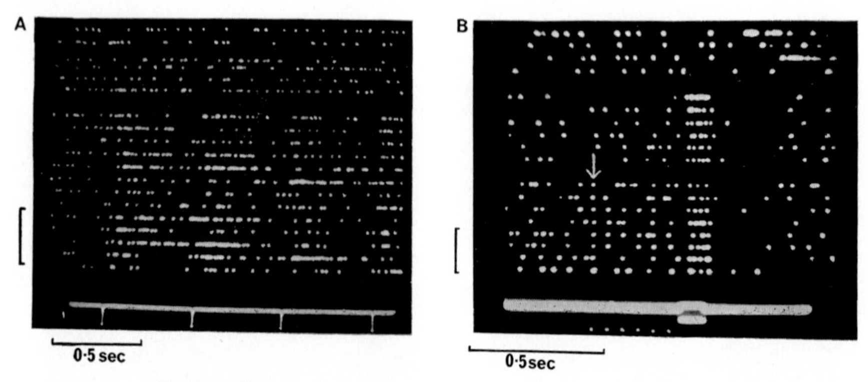

FIG. 13 (A) Hit-and-miss responses of an audio-visual unit to clicks at intervals of 0·5 sec. This unit gave responses of 0·1 to 0·5 sec duration to single clicks without habituation at 4 sec intervals. (B) Test for anticipatory enhancement. The first part of the record shows the background discharge; the middle part shows responses of an audio-visual unit to small flashes alone; the third block of lines shows responses to a series of six clicks which now precede the same flash. The single-spike responses to the first click are shown by the arrow. The response to the flash is slightly reduced (from 5·5 to 4·1 impulses per flash) by the clicks, and, despite numerous repetitions, no example of progressive changes suggesting association was ever found.

The usual effect of superimposing the two responses (taking account of the latency differences between modalities by starting the visual stimulus earlier) was to increase the number of impulses although the total was less than the sum of two separate unimodal stimulations (Fig. 13B).

The effect of a sound pulse before a flash was tested on most units to observe any possible change in the subsequent response to the flash after many repetitions, with the stimuli at sufficiently long intervals for the responses not to run together. No significant long-term effects appeared. This result was similar for the converse situation, flash preceding sound, except that light stimuli frequently have a post-stimulus period of inhibition, during which the responsiveness to sound is reduced.

F. *Latency*

Values of 30 to 200 msec were typical for visual responses. Audio responses had latencies of 50 to 70 msec and some units showed a longer latency to light than to sound. Measurements of latency have been used as an indication of how many synapses have been crossed on the path to the unit from which the record is made (Burtt and Catton, 1959). However, repetition of the stimulus and the method of representation by successive lines of dots shows that latencies are by no means constant. Changes in latency are in fact inevitable if any intervening unit has a background discharge or if an inhibitory process is involved. The occurrence of two linked spontaneous variations in latency might be used to demonstrate a truly causal interaction between neurones, as distinct from independently derived effects of the same input.

G. *Patterns of Spikes*

As for latencies, the exact patterns of spikes in responses are not consistent on repetition. Examination reveals all types of irregularities which may be worth detailed examination, if only one knew what features may be significant. At intervals there are large bursts of spikes in most of the higher-order wide-field units and multimodality units. Sometimes such bursts are obviously associated with chewing or struggling movements but frequently there is no obvious cause.

In the brain there are numerous large M units with a regular background discharge of 1 to 5/sec which cannot be modified by any reasonable stimulus. At irregular intervals these units have bursts at frequencies of 50–150/sec. The incidence and duration of the bursts have not been related to any other changes, although it is probable that some of the unaccountable bursts in the auditory and visual interneurones have their origin in these strange central units.

Some visual units give an oscillatory response when a fairly strong stimulus is switched on and held on. There is an initial burst at "on", followed by a pause of 50 to 500 msec which is followed by a second burst or series of bursts (Fig. 14). The explanation seems to lie in the existence of short-lived but long latency inhibitory effects somewhere upstream from the unit which gives the response.

H. *Novelty Units*

These are distinguished by high sensitivity to an initial movement, rapid habituation to the same movement when repeated, but showing approximately the original sensitivity to another movement *somewhere else* in the visual field (Fig. 15). The field size is usually the whole eye and commonly both eyes. When these units were first encountered we tried a great number of experiments with spatially separated movements

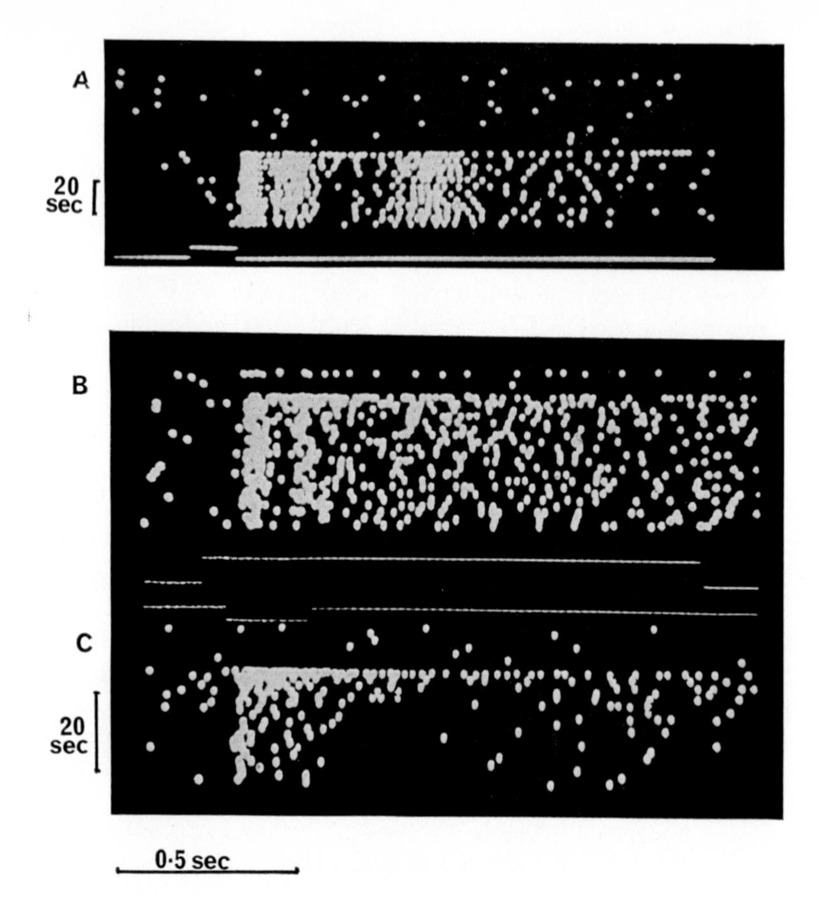

FIG. 14. Audio-visual unit from optic medulla region. Each dot represents a nerve impulse and the picture reads like print on a page. The first few lines of each record show the background discharge in the absence of a stimulus. (A) Responses to a contralateral small flash from a DM 160 at intervals of 4 sec. There is a peculiar double bounce at about 0·1 sec and 0·5 sec after the signal. (B) Response to a long-maintained ipselateral light flash with other conditions remaining the same. The second bounce is no longer there, showing its origin to lie in the "off" part of the stimulus, but the first bounce is clearly a persistent feature. (C) Response to a contralateral sound pulse of 10 kc/s for 0·25 sec. Although in the optic lobe, the latency to light (90 msec) is greater than to sound (20 msec). The latency and the form of the habituation is quite different from that with a light stimulus, and there is no bounce. Sound responses of this unit are illustrated in Fig. 7.

FIG. 15 (*facing*). Novelty unit (audio-visual). Responses to stimulation of only a few ommatidia with repeated small light flashes which were delivered from a 10μ diameter light guide applied to the ipselateral eye surface. Background dark-discharge is shown on the first halves of (A), (B) and (E). Stimuli are monitored by the continuous traces. (A) The initial large "on" and small "off" responses adapt with repeated presentation, the "off" component disappearing more rapidly. (B) Like (A), but stimulation of the contralateral eye. The habituation shows a slightly different form. (C) Repeated ipselateral presentation

[*Continued on page 191*

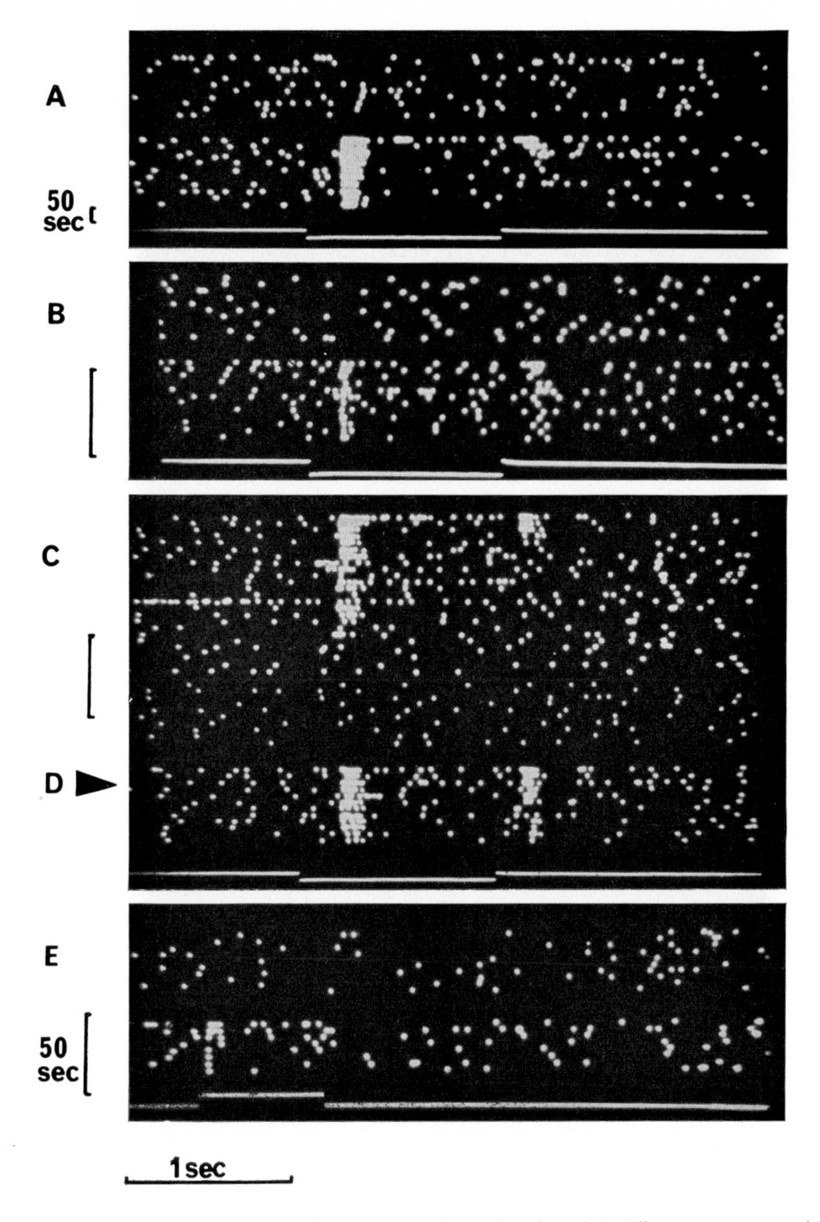

eventually produces complete adaptation of both "on" and "off" components, with differing time-course. (D) At the point shown by the arrow, the light guide was moved a few ommatidia in between flashes, whereupon the responses reappeared. (E) The unit responded feebly to sound (10 kcs, ipselateral) with quite a different form of habituation. This unit acted as a novelty unit with respect to movement in different areas of the visual field; it showed no dishabituation phenomena of one stimulus by another; it also responded at "off" to light-guide stimulation of the ipselateral ocellus. Responses to one type of stimulus were unaffected by habituation to a different stimulus.

of different types such as rotation and oscillation in a search for interaction, but without finding any features of interest. They are not sensitive to particular directions of movement, do not discriminate against changes of brightness or stationary flashing lights, and the stimulus could well be any small local change in brightness. The field over which the habituation is effective is a little larger than the area of the actual stimulus, so that these units do not allow a useful measurement of acuity. Habituation to one type of movement includes habituation to other types of movement in the same area, so that, for example, a linear movement is not distinguished from a rotary movement which it may cross. No arousal or dishabituation has been recorded in novelty units. The time for recovery from habituation can be several minutes and bears no relation to the time for habituation or to receptor adaptation. The characteristics of the habituation rates to different types of stimuli at various repeat intervals vary within a small range for different units, or the same units in different individuals, but these features cannot be modified in a given unit.

I. *Movement Receptor Units*

The existence of optomotor responses shows that the direction and velocity of movement of the whole visual field can be accurately measured by the visual system. Units which respond to movements of contrasting objects in particular directions can be found in the medulla region and deeper in the optic lobe. To prove that movement is the feature of interest it is essential to eliminate intensity change as the stimulus and to have a silent motion in case the response is partly an auditory one. Actual movement of a contrasting object always includes an intensity change on some of the receptors. Two partially adequate controls are available: (a) The unit must be relatively insensitive to changes in general background illumination or to flashing in the field of the movement. (b) A one-way movement receptor is defined as showing an increase in discharge to movement in one direction but a decrease in the opposite direction together with insensitivity in the directions at right angles. These criteria exclude complex "on-off" units such as that illustrated in Fig. 16. In practice, rather than an actual movement it is convenient to use a virtual movement which is produced by an illuminated patch switched off at one point at the same instant as an identical bright patch appears at a neighbouring point, as in Fig. 17. Tuning indicators make small light sources which can be conveniently switched. Change in brightness is checked with a photocell and both lights can be switched on or off together as a control against brightness

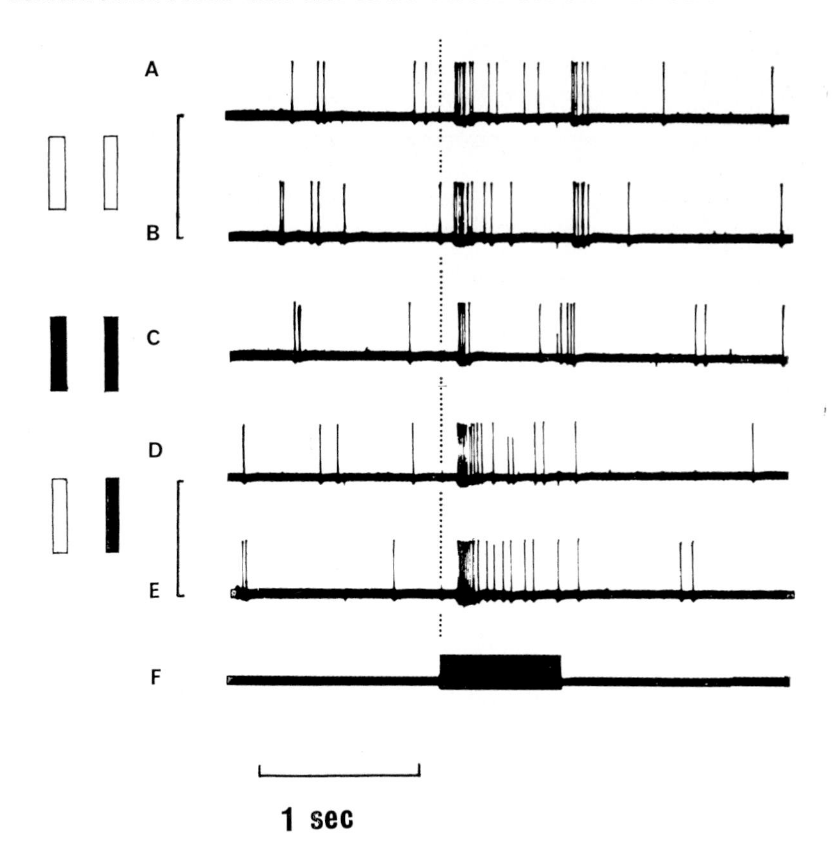

FIG. 16. Impulses of a unit giving "on" and "off" responses to single lights, but an entirely different pattern to more complex stimuli. In A and B, transient bursts of impulses arise out of a weak background discharge when two light sources separated by approximately 3 degrees in the field are simultaneously switched on and off. In C, both lights are continuously on, and briefly interrupted during the period indicated by the stimulus monitor trace (F). The responses are those to be expected. In D and E, one light, normally off, is pulsed on while the other, normally on, is simultaneously interrupted. The entirely different response pattern shows that excitation initiated at "on" and "off" is not additive. As far as could be determined, the unit had no specific sensitivity to movement, so that the new response does not represent a sensitivity to the simulated movement (phi-phenomenon) presented by the complex stimulus in D and E. Results of this kind suggest that classifications using any simple stimuli are based upon a partial knowledge, but more sophisticated properties, if discovered, may still not be those of relevance to the animal.

effects. The method, however, does not allow variation of the speed of movement, and it is possible that some units may reach saturation at slow speeds. Range fractionation (i.e. some units responding to fast, others to slow movements) is not so obvious a feature of insect visual

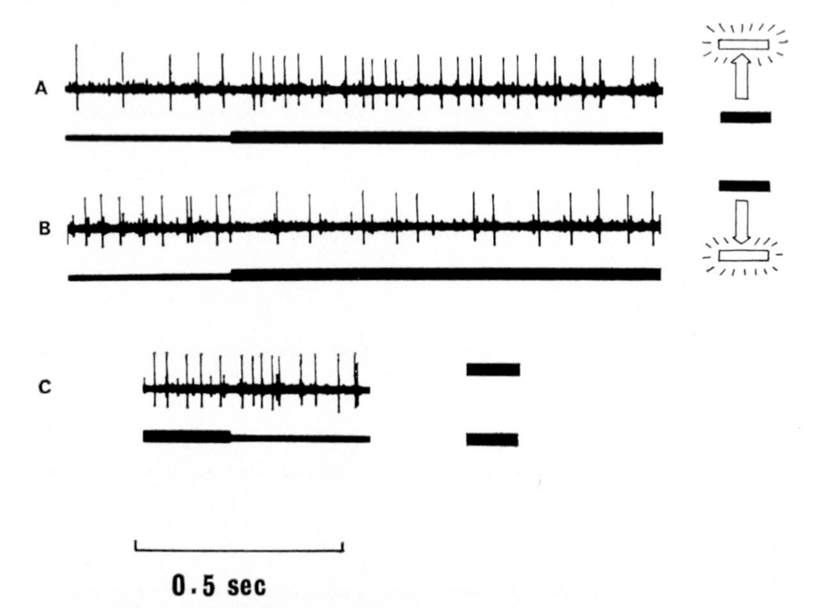

FIG. 17. The response of a one-way movement unit to simulated movement caused by instantaneously switching a light source to a new position (modified phi-phenomenon). Though the change, marked on the recordings as a thickening of the monitor trace, is instantaneous, the response may last several seconds, and is slightly better when the two positions are close together. In the recordings of this figure they are separated by ten degrees. In (C) the lights at both positions are turned off simultaneously to show the virtual absence of response in spite of the relatively large change in intensity. Phi-responses elicited in this fashion are largely independent of the position of the pair of lights in the visual field, providing their orientation is constant.

movement receptors as it appears to be in Crustacea (Waterman, Wiersma and Bush, 1964). Responses from such a movement receptor unit are shown in Fig. 6. There is a wide receptive field of some sixty degrees, and a maintained discharge that is influenced by movement of contours parallel to a single axis, in this case the vertical one. The stripes subtend about five degrees to the animal, and it is arranged that their movement will not give rise to cycles of net illumination change. Movement of the pattern in one direction (upward) accelerated the discharge rate, and when movement stops the response shows an inertia carrying on for a few seconds. This continuation of the response for a time suggests that in this case the autocorrelation of the stimulus on one area of the eye, with the stimulus on an adjacent area, involves a mechanism with delays of several seconds. Movement in the opposite direction virtually arrests the maintained discharge, and this inhibition shows a complementary inertia at the end of stimulation. Slow sustained movement was the most effective stimulus for this particular unit, which saturated at angular velocities of 2°/sec.

Since the discharge carries quantitative information about several parameters of the effective stimulus, such as the extent and contrast of the contour, its speed of movement, and the extent to which its orientation conforms to the ideal, the unit can be said only with reservation to classify unambiguously the rather abstract quality, directional movement.

Responses to virtual movement depend to some extent on the distance of separation of the two lights: the closer they are, the greater the response. Such a response depends entirely on the correct orientation of the lights: if they are at right angles to the plane of sensitivity of the unit there is no perceptible change in the maintained discharge. If conditions are correct, while no net illumination change has occurred in the receptive field, an entirely obvious and sustained response can be produced. On the other hand the unit is an insensitive detector of localized illumination steps. For a relatively much larger and diffuse light source it conforms to the classical description of "on-off" units, the responses being entirely transient. Using a small light source responses were, as far as could be determined, equivalent throughout the field.

A similar type that has been seen is one with a smaller field, responsive optimally to centrifugal movement and inhibited by centripetal movement; most of the above comments apply to this type.

A search has been made, without success, in the brain, neck connectives and ventral cord, for movement receptor units without habituation because their presence is suggested by the optomotor responses.

J. *Arousal or Dishabituation*

Habituation rates of units differ markedly, and clearly do not depend on receptor adaptation. No functional significance for the variety of habituation rates has been suggested by any of our results. The existence of neighbouring units which habituate only at higher frequencies of the same stimulus shows that the excitation certainly penetrates. We investigated the possibility that after habituation the response may be dishabituated at the original latency by means of an additional stimulus. However, despite a careful search, only rarely was there any sign of arousal. The only substantiated case was an arousal of a habituated simple "on" response to a light flash when the new stimulus, a sound pulse, shortly preceded the light flash (Fig. 18).

III. Discussion

A relatively small number of units are encountered in any particular location, although nerve fibres are known to be crowded closely. This situation is peculiar when considered with the fact that some units can be recorded over distances that are large compared with the size of the axons, as illustrated in Fig. 19. Electron micrographs show that the majority of the nerve fibres in the ganglia are in the size range of less than 2·0μ, while experience with indium microelectrodes, coupled with examination of sections stained specifically for nerves, suggests that recordings are possible from fibres in the size range 2–5μ diameter. We conclude, therefore, that the relatively large silent zones and paucity of units reflect the inadequacy of the electrodes to pick up small units and cannot be held as evidence of a lack of responses.

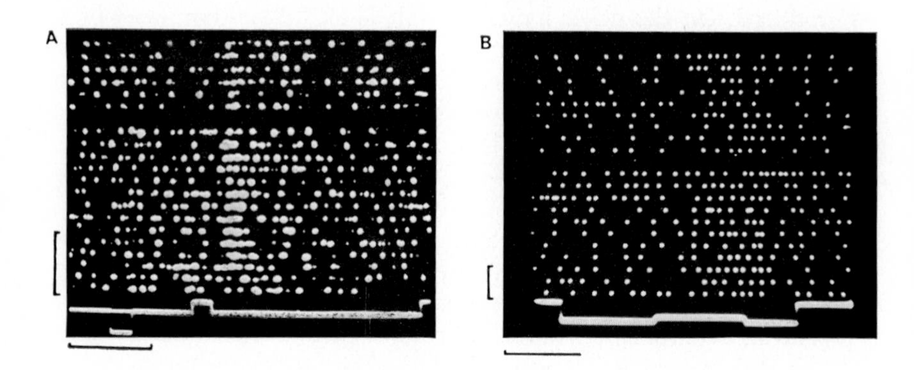

Time marks = 0·5 sec (horizontal)

12 sec (vertical)

FIG. 18. Arousal of the habituated response by a stimulus of a different type. (A) the first few lines, down to the omitted line, show the partially habituated responses to a small ipselateral light flash which is repeated every 2·5 sec, as indicated by the upwardly directed pulse on the stimulus trace. In the lower part of the record an ipselateral sound pulse of 10 kc/s, as indicated by the downwardly directed stimulus mark, precedes each light pulse by 0·5 sec. The result is a negligible response to the sound but an increase in the visual response at the same latency as before. This, in turn, will eventually habituate. The interval between the two stimuli is critical although the nature of the sound and light stimuli seems to be of no importance. This phenomenon was seen in only one unit but on that occasion it was substantiated many times with a variety of sound and light stimuli. (B) The upper part shows the response to an ipselateral 0·5 sec long note of 10 kc/s as indicated by the upwardly directed pulse in the stimulus trace. The lower lines show the slightly increased response when the sound pulse is accompanied by a small light stimulus as shown by the downwardly directed long pulse. This illustrates how, in this case, the arousal effect could be caused by a small change in background discharge.

Perhaps the most disappointing failure of the technique is the inability to identify pathways of excitation from second order to third to fourth order interneurones. On account of their variability, latencies have proved to be a poor indication of the number of synapses removed from the stimulus. Latencies do not indicate causal chains, but only temporal relations of events which may be only indirectly casually connected, because so many parallel lines are active. Under typical stimulating conditions almost any two units show a marked correlation. In order to achieve an advance in the significance of the results, it

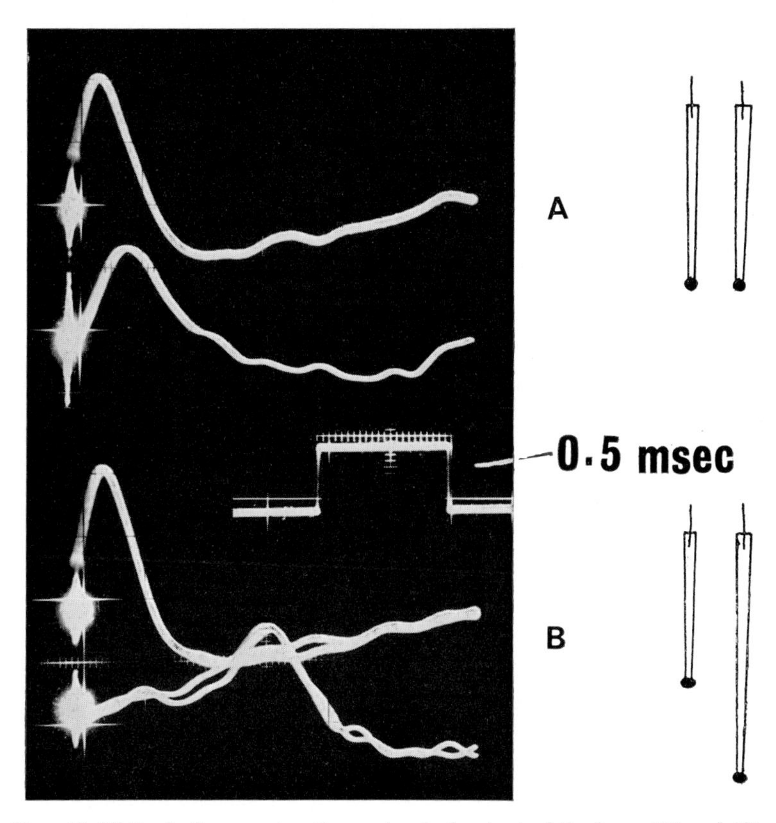

FIG. 19. Method of measuring the anatomical extent of the large (A) and (C) units. (A) Two electrodes are advanced parallel with one another and record spikes of a large unit in the medulla. Spontaneous spikes recorded by one electrode trigger the oscilloscope sweep, showing the constant delay before the spike arrives at the other electrode. This eliminates responses from all other units which may interfere. (B) An advance of one electrode by a distance of 200μ causes a lag due to conduction velocity. The use of two electrodes eliminates the effect of movement of the tissue by the advance of a single electrode, and permits the certain identification of the same unit at two different locations.

appears that multichannel recording from a much higher percentage of histologically identified units is necessary. Even then, the signs are that much of the observable activity will have to be attributed to causes in a large unseen population of small neurones. This is a fundamentally unsatisfactory situation when the aim is the analysis of mechanisms, and not just the collection of a variety of neurone types.

A. *Types of Stimuli Used in Analysis*

Controlled stimuli are essential for analysis and the results obtained with them can be related to the performance of models. However, the choice of stimulus, whether flashing light, moving light, moving black-white contour, diffuse background light or a more complex pattern, depends on what types of normal stimuli are found by trial to be effective. If the animal has units which are highly specialized to respond selectively to certain aspects of the visual field of its particular environment then it is not easy to discover exactly what these aspects are. In such circumstances readily controlled stimuli may produce "spurious" responses at thresholds that are high relative to that of the optimal natural stimulus. Abnormal stimuli may do no more than reveal imperfections to the gating mechanisms which allow these units to classify more complex situations.

B. *The Ambiguity of Units*

If an animal is to derive information from its neurones one might suppose that their responses should represent unambiguously the stimuli which fall upon the sensory cells. However, clearly this does not happen. Every unit which we have examined is sensitive to a variety of attributes of the stimulus, so that the quantitative, temporal or patterned characteristics of the stimulus are not recoverable from spike patterns as seen in this unit alone. There have been no truly specific recognition units for complex stimuli among hundreds of units tested. Units which record preferentially to general attributes such as movement, or to two clicks at a certain interval, always respond to some stimuli which do not have these features, even though the threshold is higher. Thus, our best examples of a generality unit, the one-way movement receptors, show ambiguities in being influenced by factors such as stripe width, speed of movement, black-white contrast, length of stripes, direction of contrasting edge in relation to direction of movement, and straightness of edges.

The ambiguity is presumably resolved further down the line in cases where the signals are of survival value. Numerous parallel pathways are always excited. In the case of one-way movement detectors, which are numerous in insect and crustacean optic tracts (Wiersma, Waterman and Bush, 1964), one final behavioural resolution, the optomotor response, cannot be elicited by any stimulus except movement of the visual field, although other stimuli may excite movement detector units. In the crab optokinetic response the afferent stimuli are centrally summed to give a centrally initiated efferent signal which very effectively controls the eyestalk movement without proprioceptive feedback of the eye position (Horridge and Sandeman, 1964). One can reasonably suppose that many afferent paths are summed, so leading to an increase in accuracy, but in this particular example a survey reveals only the motorneurones as standing in the position of final integrating units.

C. *The Basis of Naming Units*

Naming of units depends on the mental predisposition to make tests with certain classes of stimuli, and on the availability of stimuli. Theory plays a large part in deciding what tests to use. The present work has been much influenced by the analysis of the insect optomotor response (Hassenstein and Reichardt, 1959; Reichardt, 1962), by work on unit activity in vertebrate brains (Maturana *et al.*, 1960; Jacobson and Gaze, 1964), on crustacean optic tract units (Waterman *et al.*, 1964) and on earlier work with locusts (Burtt and Catton, 1960). Agreement with theory itself causes a tendency not to look further; for example, the one-way movement detectors may yet prove to serve quite a different function. Therefore a re-examination of the properties of the units is required whenever a new idea comes to the investigator. An example is the discovery that the net convexity (small black spot) detectors of Maturana *et al.* (1960) in fact have a field with inhibitory centre and excitatory surround and are therefore not stimulated by a contrasting straight boundary which would cut across the centre (Gaze and Jacobson, 1963). Mapping with small light spots gives one type of result with early-order visual units, but higher-order units which may be gated against such stimuli cannot be expected to reveal their preferred pattern.

D. *The Necessity for a Behavioural Correlate*

The number of neurones and the variety of the possible stimuli are so great that a sense of proportion must be maintained. The number of possible interactions between neurones is also so great that the analysis of which neurones act on which others must follow a plan which is related to the animal. A behavioural correlate can supply a guide for the analysis. The dependence of the optomotor response on one-way movement detectors is one example of this. The sensitivity of the stickleback or robin to red, or the frog to blue light, provide other examples where the criterion of a useful observation in the brain would be its relation to the quantitative features of the whole response. It would have been possible in principle to determine the spectral characteristics of all the visual units reported in the locust. This might have led to a conclusion that there is a diversity of spectral response in the primary receptors. This, however, is already known to be the case in several insects by direct recording from these receptors.

A general conclusion from this exploratory investigation is that the responses of individual units show great ambiguity and little sophistication of analysis. However, numerous units are active at the same time. Therefore, in simple reflexes, such as startle responses, and optomotor responses, the final behavioural output by the motorneurones represents the summation of a great deal of simultaneous, relatively simple excitation of interneurones in parallel. About more complex acts of behaviour, such as recognition of pine cones around a nest-entrance, nothing can be said.

IV. Summary

1. Extracellular recordings reveal a variety of auditory and visual responses of single units in different areas of the brain and optic lobes of the migratory locust. About 25 types of unit are characterized, mainly on responses to light flashes and movement.
2. Habituation is typical on repetition of the stimulus; rate of onset and recovery are varied. Where a unit has many inputs, as in audio-visual or wide field visual units, habituation at one input usually leaves responses to other inputs unchanged.
3. Among visual units, one-way movement detectors, ones with excitatory centre and inhibitory surround and ones with inhibitory centre and excitatory surround demonstrated the partial abstraction of a certain characteristic of the stimulus. Most units are rather non-specific in their responses, and in none can the nature of the stimulus be inferred unambiguously from the responses.
4. There are numerous large widely ramifying non-specific units which respond to any local brightness change on either eye and to many mechanical stimuli including sound.
5. A small degree of lateral inhibition can be demonstrated in certain units.

6. Numerous efforts to demonstrate sensory-sensory associations by repetition of paired stimuli always led to no result.
7. Dishabituation of a response to a visual stimulus can be brought about by a preceding sound stimulus, but units showing this are rare.
8. Recordings show that any stimulus will cause activity in numerous parallel lines in the nervous system.
9. Limitations of technique mean that only a small fraction of the neurones can be picked up. Presumably the specific responses of neuropile areas depend on local small neurones and these cannot be recorded.
10. The following methodological problems arise: (a) ignorance of the normal or even of the optimum stimulus, (b) inability to disentangle the interactions of diverse and unidentified units, (c) the significance of sequences of impulses which may well be ignored at the next stage down the line, (d) how to do more than collect a few inexplicable phenomena when most of the activity is in neurones below the size which can be recorded, (e) the importance of studying the responses of neurones (apart from early order sensory ones) in relation to a normal behavioural response which can give meaning to the whole exercise.

REFERENCES

Burtt, E. T., and Catton, W. T. (1954). Electrical responses to visual stimulation in the optic lobes of the locust and certain other insects. *J. Physiol.* **133,** 68–88.

Burtt, E. T., and Catton, W. T. (1959). Transmission of visual responses in the nervous system of the locust. *J. Physiol.* **146,** 492–515.

Burtt, E. T., and Catton, W. T. (1960). The properties of single-unit discharges in the optic lobe of the locust. *J. Physiol.* **154,** 479–490.

Cajal, S. R., and Sánchez, D. (1915). Contribución al conocimiento de los centros nerviosos de los insectos. *Trab. Lab. Invest. biol. Univ. Madr.* **13,** 1–164.

Gaze, R. M., and Jacobson, M. (1963). "Convexity detector" in the frog's visual system. *J. Physiol.* **169,** 1.

Gesteland, R. C., Lettvin, J. Y., Pitts, W. H., and Rojas, A. (1963). Odor specificities of the frog's olfactory receptors. *In* "Olfaction and Taste" (Y. Zotterman, ed.). Pergamon Press, Oxford.

Hassenstein, B., and Reichardt, W. (1959). Wie sehen Insekten Bewegungen. *Umschau* **10,** 302–305.

Horridge, G. A. (1961). Pitch discrimination in locusts. *Proc. roy. Soc.* B. **155,** 218–231.

Horridge, G. A., and Sandeman, D. C. (1964). Nervous control of optokinetic responses in the crab *Carcinus*. *Proc. roy. Soc.* B. **161,** 216–246.

Hoyle, G. (1955). Functioning of the insect ocellar nerve. *J. exp. Biol.* **32,** 397–407.

Huber, F. (1962). Central nervous control of sound production in crickets. *Evolution* **16,** 429–442.

Jacobson, M., and Gaze, R. M. (1964). Types of visual response from single units in the optic tectum and optic nerve of the goldfish. *Quart. J. exp. Physiol.* **49**, 199–209.

Maturana, H. R., Lettvin, J. Y., McCulloch, W. S., and Pitts, W. H. (1960). Anatomy and physiology of vision in the frog (*Rana pipiens*). *J. gen. Physiol.* **43**, Suppl. 129–175.

Reichardt, W. (1962). Nervous integration in the facet eye. *Biophys. J.* **2**, 121–143.

Scholes, J. H. (1964). Discrete subthreshold potentials from the dimly lit insect eye. *Nature, Lond.* **202**, 572–573.

Suga, N., and Katsuki, Y. (1961). Pharmacological studies on the auditory synapses in a grasshopper. *J. exp. Biol.* **38**, 759–770.

Varjú, D. (1959). Optomotorische Reaktionen auf die Bewegung periodischer Helligkeitsmuster. *Z. Naturf.* **14b**, 724–735.

Waterman, T. H., Wiersma, G., and Bush, B. M. H. (1964). Afferent visual responses in the optic nerve of the crab, *Podophthalmus*. *J. cell. comp. Physiol.* **63**, 135–155.

Neurophysiological Studies on "Learning" in Headless Insects*

GRAHAM HOYLE

Biology Department, University of Oregon, Eugene, Oregon, U.S.A.

I. INTRODUCTION

The unknown mechanisms involved in the learning process present some of the major unsolved physiological problems. One reason for this is that most of our knowledge about the grosser aspects of learning has been achieved in birds and mammals whose nervous systems are so complex that they defy analysis at the unit level. In an attempt to provide an example of learning in an experimentally accessible situation, namely a thoracic ganglion of an insect, Horridge (1962) presented headless cockroaches (*Periplaneta americana*) and locusts (*Schistocerca gregaria*) with a simple postural learning problem: to maintain a leg in the raised position or to receive an electric shock. The insects were placed over dishes containing saline carrying a lead through which electric pulses could be delivered from a stimulator. The leg itself carried a wire connected to the other terminal of the stimulator so that the circuit was completed every time the leg touched the saline.

The legs always extend spontaneously, but after a few shocks they may flex and remain flexed so that shocks are not received. If the receipt of a shock is termed an error, and the number of errors per unit time for several animals are averaged and plotted against time, then it is seen that there is a marked progressive decline in the number of errors occurring over a period of about 10 minutes. The headless preparation may be said to have "learned" to avoid being shocked. This requires the maintenance of an unnatural posture. Lowering the animal can result in its receiving shocks again, but this is followed by a raising of the leg still higher. I have repeated some of Horridge's experiments and have confirmed the results in a general way. A detailed repetition of the work has been made by Cohen and Eisenstein (1965). One difficulty is in the great variability experienced between animals. Some never lift the leg out, but respond by even more severe extension so

* Supported by research grant G-21451 from U.S. National Science Foundation.

that they receive shocks continually. A few receive but one shock, lift the leg and then do not extend it again for hours. The majority experience many shocks at progressively less frequent intervals interspersed with periods of relatively prolonged extension. These variations must be explained and made compatible with any neurophysiological hypothesis of learning mechanism.

II. Preliminary Studies

As a preliminary to neurophysiological examination of the "learning" mechanism several points not considered by Horridge needed elaboration. For instance, which muscles are concerned, and in what way? Extension of the leg may be the result of either an increased extensor discharge in one or more of several extensor muscles or a decrease in discharge of one or more flexor muscles, or both may occur simultaneously. The muscles which might be concerned are as follows.

Extensors: extensor tibiae, extensor trochanteris (five muscles), coxal abductors (three muscles), tergal promotor, coxal promotor.

Flexors: flexor tibiae, flexor trochanteris (two muscles), coxal adductors (two muscles), coxal remotors (two muscles), tergal remotors (two muscles).

In the cockroach it seems that the majority of preparations use flexion at the trochantero-femoral joint to raise the leg, but they may also use a twisting upwards and backwards of the whole coxa. If the femur is parallel with the water, then either extreme flexion or extreme extension at the femoro-tibial joint will lead to avoidance. Either may be used in long-term raising of the leg (Cohen and Eisenstein, 1965). In the locusts studied some animals showed marked flexion at the trochantero-femoral joint after a few shocks and some showed levation of the trochanter. Neither effect persisted for a long time. Recording of electromyograms with pairs of fine leads showed that these effects were associated with an increased background discharge in the appropriate flexor muscle. More persistent leg-raising was achieved by an inward and upward movement of the whole coxal joint. This was associated with a prolonged electrical discharge which could be recorded best from just above the coxa. It seemed probable, then, that an increased discharge in a coxal levator muscle most commonly caused the avoidance reaction, possibly coupled with inhibition of the tendency to fire spontaneous bursts in extensor muscles.

If the learning process really achieves a prolonged discharge, then this might still be present after further dissection of the whole animal.

Experiments with headless cockroaches trained to hold up one leg failed to reveal such discharges, but in locusts it was found that those which had been trained consistently showed a high-frequency maintained discharge in the two coxal adductor muscles. No other muscles which could have been associated with leg-raising showed a similar prolonged discharge consistently. It has since been found that those muscles show a spontaneous background discharge at all times. However, the high frequencies encountered after training are uncommon in fresh preparations after due time has elapsed since decapitation (see below). The adductor muscles, then, provide a base for the experimental study of the learning process.

They are attached, one above the other, to the posterior margin of the meso-sternal apophysis on their inner borders, and to the posterior inner rim of the coxa (mesothoracic M100), or the posterior angle of the coxa (mesothoracic M101) according to the designation used by Campbell 1961 on their outer borders. Since their inner attachments are set higher than those on the coxal rim their action is to pull the rear, inner rim of the coxa inwards towards the body and slightly upwards. The concerted action of these muscles thus raises the whole leg. The leg-raising is much more pronounced when the tergal remotor (mesothoracic M90) is active, or the levator trochanteris. However, neither of these muscles has been found to give prolonged discharges such as would be required for maintained leg-raising.

The more ventral of the two coxal adductor muscles is attached to the centre of the coxal rim and is therefore anterior to the other adductor. It will henceforth be referred to as the anterior coxal adductor (a.c.a.) and the more dorsal the posterior coxal adductor (p.c.a.). The abbreviation mt. will be used to designate metathoracic and ms. mesothoracic.

In order to establish the necessity of correlation of the shock with leg position for achieving learning, Horridge used a second (R) animal linked in parallel with the first (P) which received shocks at the same time as the first, but in its case the shocks were not correlated with leg position. Not only did these animals not raise their legs, but on subsequent tests R animals took longer to learn in the P (correlated) situation or did not learn at all (i.e. an exposure to shocks not correlated with position reduced the capacity to learn an association subsequently).

The type of tests made by Horridge are, however, not critical as to the significance of leg position. The feature which must lead to a shock being received, as pointed out above, is either an extensor thrust, a fall in flexor discharge frequency or a combination of the two, although the effects of these could be monitored via the afferent discharge associated with leg movement. It may be that the arrival of a burst of impulses in sensory nerves from the leg following the significant change is reflexly followed by a compensatory change in the nervous system so as to reduce the probability of recurrence of the same change. Such a change need not require specific afference from limb proprioceptors indicating position. Pursuing this line of argument, the last sentence may be framed in the form of a hypothesis and put to the test.

The hypothesis is that if, following a sharp change in frequency in a motorneurone whose continuous output is associated with significant posture, a burst of sensory impulses occurs in the nerves of the same limb, then the frequency of that motor discharge will undergo a prolonged alteration in the direction counter to the change.

It would not be surprising if such a mechanism existed and were a general one, for it would have obvious adaptive significance. If it did exist, and had been known, then the experience of Horridge on headless insects could have been predicted. The apparent immediate adaptation of the insect locomotory control mechanism to such drastic measures as amputation of a limb might be explained on the same basis.

III. Experimental Methods

The animals studied were: *Schistocerca gregaria*, *Melanoplus differentialis* and *Periplaneta americana*. They were secured, back down, in soft dental wax and decapitated. Then 0·007″ silver wire was wrapped around the mid-tibia of from one to four legs, according to the experiment, in about three close turns, and taken to a terminal. A single loop of similar wire was tied round the middle of the tarsus and also connected to a terminal. A Grass S 4 stimulator was connected to the terminal for a given leg.

The loose connective between the rim of the coxa and the thorax was then cut away so as to expose the underlying coxal and trochanteral muscles.

For examination of neuromuscular transmission the motor nerve to the muscle to be studied was sectioned and placed on tapered, chlorided silver wire electrodes for stimulation. The apodeme was cut at its distal insertion and seized by a micro-forceps mechano-electronic force transducer. The tip of a glass capillary micro-electrode filled with 3 M KCl was then brought close to a muscle and intracellular penetration made. Ground return was via a pool of saline making contact with the open neck. In some experiments, particularly in connection with tests of transfer of learning, two muscles were exposed and examined simultaneously.

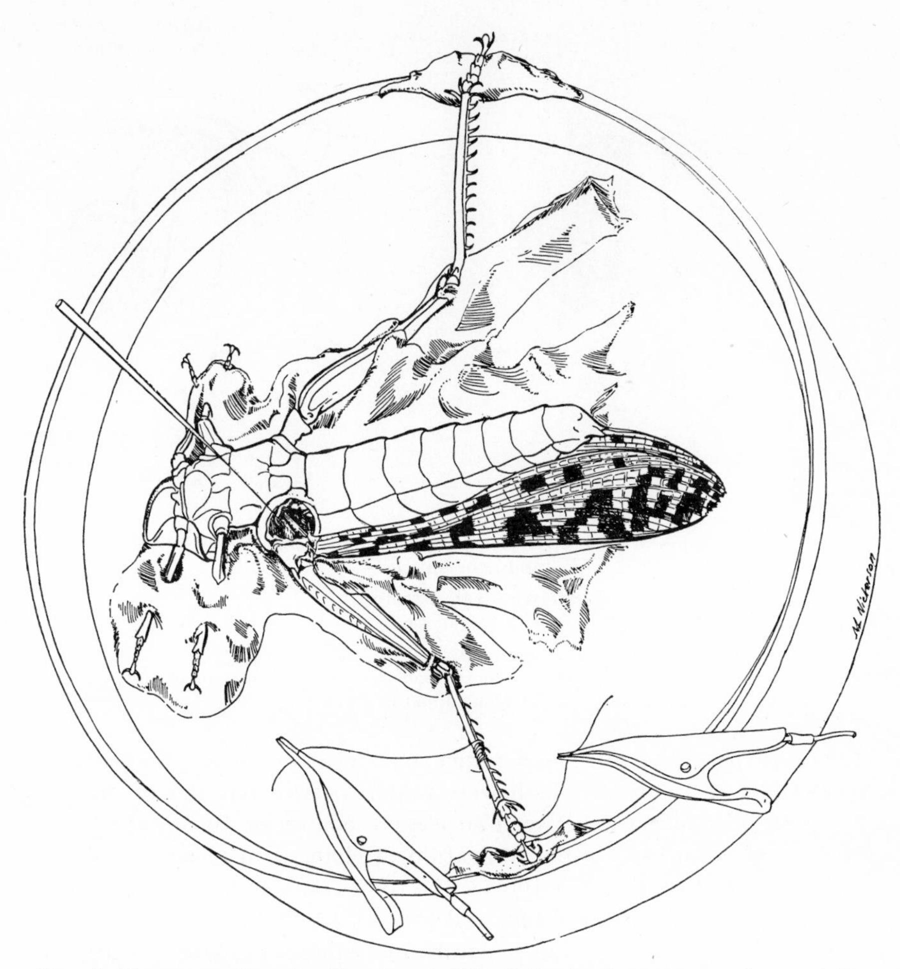

FIG. 1. The preparation—headless locust. Stimulating leads are wrapped around the tibia and tarsus. An intracellular electrode records from the exposed anterior coxal adductor of the metathoracic leg.

The output from the microelectrodes and transducer were taken through cathode-followers to a multiple-beam cathode-ray oscilloscope, 6-channel pen oscillograph and 4-channel tape-recorder (for general storage and play-back analysis). All were direct-coupled. A brief review of the instrumentation has been given elsewhere (Hoyle, 1964). The microelectrode output was also taken through pulse shapers, giving uniform, brief pulses for each post-synaptic potential going in. This instrument could resolve the minute saw-tooth effect produced by very close post-synaptic potentials occurring at frequencies up to about

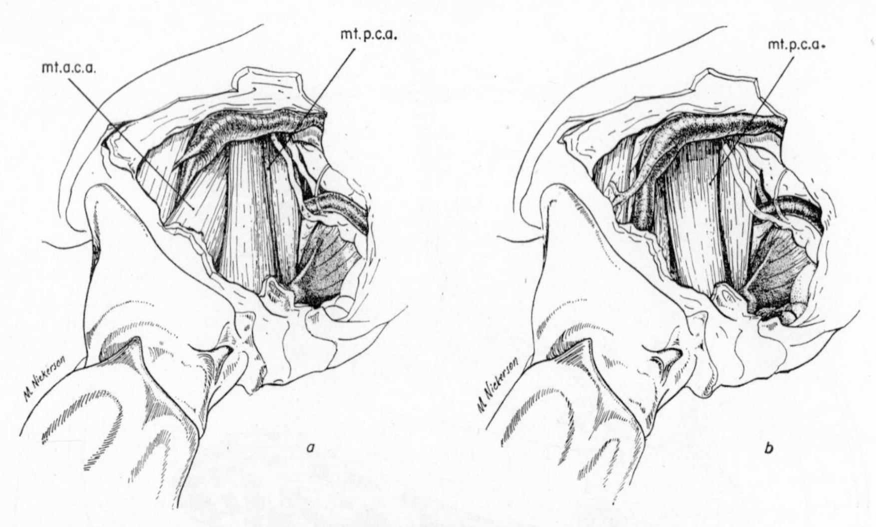

Fig. 2. Enlarged view of the base of the right metathoracic coxa after exposure to prepare adductor muscles.

mt. a.c.a. = metathoracic anterior coxal adductor
mt. p.c.a. = ,, posterior ,, ,,

(a) after removing only the integument.
(b) after removing the anterior adductor and the posterior rotator of the coxa in order to expose fully the posterior adductor.

150/sec. The pulses were next taken to a decatron scaler, the outputs from the first stages of which were taken to two channels of the pen recorder. The interval between successive counts of 10 impulses thus monitored, afforded a rapid indication of changes in mean frequency. At intervals of every 10 or 100 sec the total count in two channels was printed out onto a paper strip and the print-out time marked on the main recorder chart so that accurate matching was achieved. In this way the mean frequency was continually monitored. Loud-speakers served as additional monitors of the output.

IV. Results

All the coxal and trochanteral muscles accessible to microelectrodes from the partially dissected outside of the animal have been subjected to preliminary studies, and it is expected that several of them will show the phenomena to be described. However, it soon became apparent that the cockroach gives very inconsistent results and it was studied little. Inhibitory effects were obtained in all three animals but were most readily elicited in *Melanoplus*. By far the most satisfactory experimental animal for testing the hypothesis was *Schistocerca*. In this animal it was observed that the coxal adductor muscles always had a background discharge. In other near-by muscles there was either no background or else it was erratic. The coxal adductor muscles can readily be exposed without doing any significant damage to the preparation. Additional dissection enables one to expose the relevant ganglion at the same time.

In the metathoracic anterior adductors of *S. gregaria* (mt. a.c.a.) the excitatory discharge was found to be due to a single axon, therefore providing an excellent preparation for the experiment. Most of the work to be described was done on this muscle. The question of whether or not it is responsible for the leg-raising phenomenon observed by Horridge need not further concern us.

A. *Neuromuscular Transmission to Coxal Adductor Muscles in* Schistocerca gregaria

1. *Metathoracic anterior* (*mt. a.c.a.*). The muscle receives a branch from nerve 4 containing 2 axons. One of these causes the appearance of large junctional potentials (j.p.s) in all the muscle fibres. The resting membrane potentials of fibres in this muscle are unusually low, as measured in the intact situation with the nerve firing. The value obtained was 53 ($\pm$4) mV in locust saline (Hoyle, 1953). The j.p.s ranged from 10–40 mV (the higher level is approximate, since graded secondary responses are evoked by the larger potentials). Severance of the motor nerve was followed by a slow rise in resting potential to the value commonly encountered in other skeletal muscles of locusts (60 mV). Each junctional potential is followed by a small twitch. The twitch tension was registered by severing the apodeme close to the coxal rim and clamping it in a forceps-tip transducer (Hoyle and Smyth, 1963). Twitch tension is up to 0·5 gm. The twitch duration is about 500 msec. Tension rises progressively with frequency forming a plateau with small peaks and becoming completely smooth only at 60/sec (Fig. 3). Tension continues to rise with increasing frequency of stimulation, reaching a maximum at about 200/sec. The rate of relaxation is slow for an insect skeletal muscle; isometric tension following a maximum tetanus takes 500 msec to fall to zero.

Impulses in the second axon supplying the muscle leads to the appearance of hyperpolarizing synaptic potentials. The magnitude of single hyperpolarizing potentials varies from fibre to fibre, being zero in some, but reaching 10 mV in a few. The total duration of these potentials is about 60 msec. Their detailed properties and function will be dealt with elsewhere, but it is essential to stress that they do not cause any significant reduction in the magnitude of the tension evoked

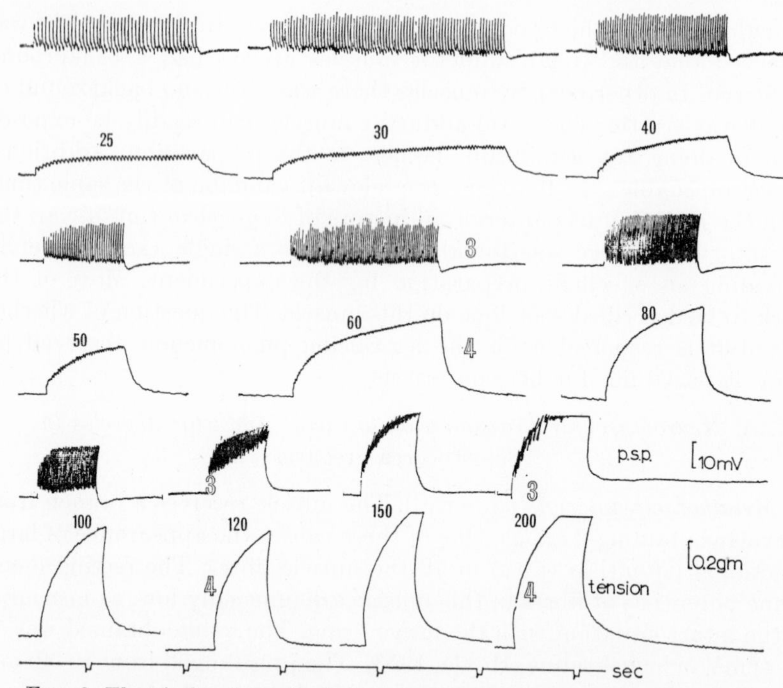

FIG. 3. Electrical and mechanical responses of semi-neuromuscular preparations. Right metathoracic anterior coxal adductor (r. mt. a.c.a.) of *Schistocerca gregaria*. Single motor axon was stimulated at frequencies stated whilst recording electrical events with an intracellular electrode from a single muscle fibre and tension from the cut apodeme of the whole muscle. Direct-coupled pen-oscillograph record. Upper traces: membrane potentials; lower traces: tension.

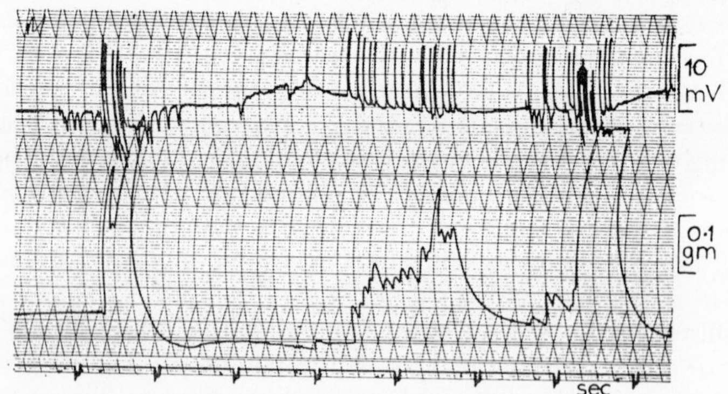

FIG. 4. Electrical (upper trace) and mechanical activity (lower trace: decreasing tension downward) recorded simultaneously from r. mt. a.c.a. of *S. gregaria* connected to ganglion during spontaneous activity. Note both depolarizing (excitatory) and hyperpolarizing synaptic potentials. When the latter are present relaxation is faster (first and third relaxations).

by stimulating the excitatory axon. The second axon gives enhancement of tension at low frequencies of excitatory stimulation, but up to 10% reduction in tension at excitatory frequencies above 10 per sec. Against a background of relatively high-frequency activity in the hyperpolarizing axon, isometric tension declines to the resting value in only 200 msec, $2\frac{1}{2}$ times as fast as normal (Fig. 4), in some preparations so that its function is paradoxical.

2. *Metathoracic posterior* (*mt. p.c.a.*). This muscle receives two excitatory axons which give j.p.s of different size (Fig. 5). All fibres in the muscle give the small size ones, but not all give the larger ones. The discharge frequency of the small ones is about 70% higher than that in the single excitatory axon supplying mt. a.c.a. (i.e. 20/sec average in the resting preparation). These discharges in the two muscles follow each other

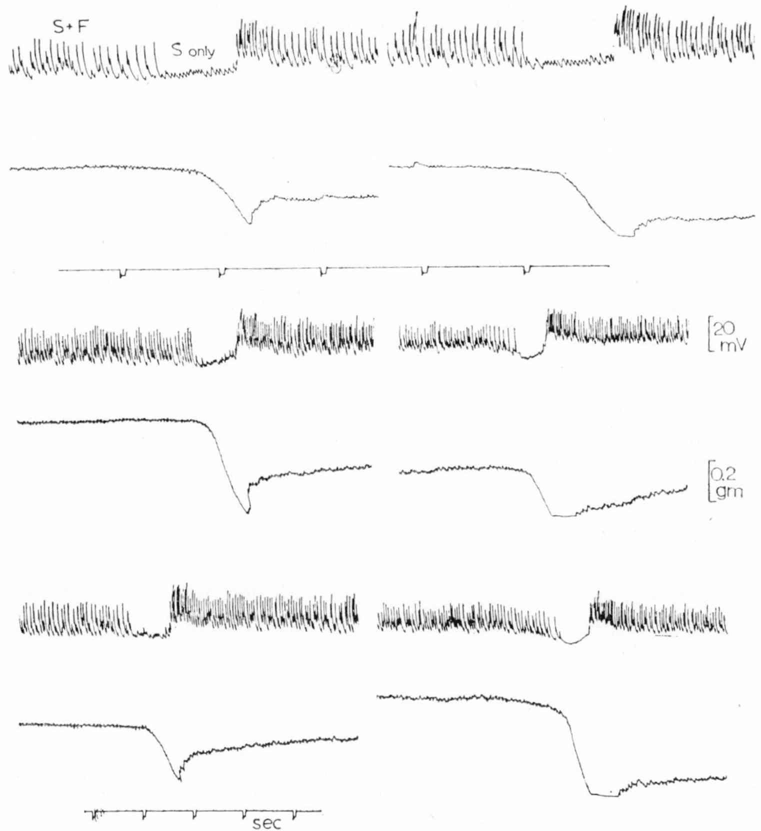

FIG. 5. Spontaneous electrical (upper trace—single fibre) and mechanical (lower trace—whole muscle) activity in right mesothoracic posterior coxal adductor of *S. gregaria* connected to ganglion. Note small and large size p.s.ps. When only large size p.s.ps. cease relaxation is partial and slow.

closely in rate. Firing of the axon giving the larger potentials is erratic, sometimes being absent, sometimes continuous. Commonly it fires in bursts which coincide with peaks in the firing rate of the other axon. In this respect it resembles the firing of the F (fast) axon supplying small extensor muscles and the other axon resembles the S (slow) axon (Hoyle, 1957).

Hyperpolarizing potentials have not been observed in this muscle.

3. *Mesothoracic anterior* (*ms. a.c.a.*) *and posterior* (*ms. p.c.a.*). The mesothoracic adductors each receive at least three axons, two of which are excitors and one a hyperpolarizer. The excitors both fire more or less continually, although bursts and prolonged interruptions are common with the axon giving the larger j.p.s (F axon). The two sizes of junctional potentials occur in the same muscle fibres. Tension falls when the discharge frequency in the F axon alone declines, but the drop is delayed and slow.

The difference between the mesothoracic muscle, which receives two excitor axons (both discharging more or less continually), and the metathoracic (which receives only one excitor), is interesting. Quantitative studies on mesothoracic adductors and the posterior metathoracic adductor are rendered virtually impossible by the existence of simultaneous dual discharges. Fortunately, however, in the mt. a.c.a. with its single excitatory axon we have an ideal muscle to study, and most of the present work has been done on it.

B. *Background Discharge*

The background excitatory discharge was recorded before decapitation in several animals. This was always about 70% higher in mt. p.c.a. (average 20/sec) than mt. a.c.a. (average 12/sec). Fluctuations were common, especially in the form of brief bursts and brief interruptions of discharge. Immediately following decapitation the rate of firing rises to 80–120/sec, being then very regular. It declines slowly, following an exponential curve which levels off after about 20 minutes (see Fig. 19). The final steady rate in the decapitated animal is always much steadier than in the intact one and is closely similar to the average in the latter (12/sec). Sometimes it will keep to within one or two pulses of its mean rate for over 30 minutes. At other times it fluctuates by up to ± 12/sec spontaneously. Background hyperpolarizing potentials were occasionally seen, and in some preparations the shocks applied to the leg elicited bursts of them. It is not considered that they had any influence on the results of present experiments. Most were done with complete absence of hyperpolarizing pulses.

C. *Response to Single, Occasional Shocks: Effect of Intensity and Duration*

The discharge in the adductors was monitored continually while single, rectangular current pulses of 2 msec duration and gradually increasing strength were applied to the leads. Below a well-marked threshold voltage no effect was observed. Above the threshold a short, reflex burst of motor activity followed. This ranged from 2 or 3 to about 30 j.p.s over and above the number to be expected from background alone and lasting as a burst up to about 3 sec. These reflex discharges were, however, extremely variable and did not always occur on repeating suprathreshold shocks. The threshold for the reflex was usually above the minimum threshold for eliciting the learning response to be described and the two are clearly related. Briefer shocks required higher voltages to be effective and no responses occurred at durations below about 0·2 msec. With progressive increases in duration above 2 msec the reflex discharge did not usually change, up to about 10 msec. Thereafter, the total number of pulses elicited increased somewhat, though seldom to more than a brief burst of about 50 pulses. With still longer shocks (100 msec and over) the reflex discharge was reduced or did not even occur, and very long pulses often caused a reduction in the rate of background firing. Experiments in which the discharge of impulses in the main leg nerves were monitored showed that increasing intensity and duration lead to both increasing frequency and total duration of impulse discharge in sensory nerves from the leg. Evidently prolonged stimulation gives rise to inhibitory effects which parallel and may overcome the excitatory ones.

Similar reductions in background rate also occurred in response to extremely strong shocks, applied at short duration.

D. *Response to Regularly Repeated Shocks*

Shocks of 2 msec duration and 10% above the threshold for eliciting a reflex burst were applied rhythmically at frequencies ranging from 1/sec to 100/sec and for various lengths of time. At the lower frequencies of stimulation, up to about 2/sec, there was a small progressive increase in the mean frequency. This increase could be attributed to the reflexly evoked bursts summating with a relatively steady background. It was not maintained on cessation of the stimulation. At the higher frequencies there was a maintained increase in background, but a rapid decline if stimulations were either stopped or continued.

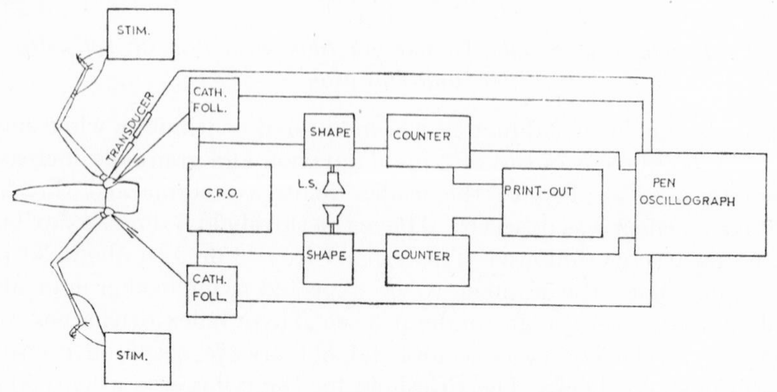

FIG. 6. Block diagram of instrumentation.

E. *Response to Selectively Timed Shocks*

If the leg-raising reaction found by Horridge is the result of an increased frequency of discharge in levator muscles, the leg-lowering will occur when the frequency drops. Since the background frequency is subject to fluctuations, we may select an arbitrary frequency value within the range of fluctuation and deliver electric shocks to the leg whenever the discharge falls below this level. Clearly this might be analogous to the actual causal phenomena occurring in Horridge's experiment. Shocks were given either singly, or at a rate of about 1/sec, following the pattern Horridge found to be successful.

The results of these tests were often dramatic. When a stimulus followed a decline in frequency the frequency almost invariably increased.

Furthermore, the duration of the increase was longer than when the same strength shock was given at a time of steady maintenance. After about 3 or 4 such shocks had been received the maintenance period was increased up to 20–30 sec or more. A minimum level of firing was set arbitrarily (I will call it the "demand" level), and the leg given a shock every time the mean frequency over a 10 sec period fell below it. After 10 or 12 shocks the level of firing was maintained at a higher value for up to 15 minutes or longer and in some experiments indefinitely. The background firing level had now been increased by 25% or more.

The above fluctuations still occurred, so the experiment was continued, raising the demand level. Again the results were similar. On the second set of shocks, the increase in frequency was relatively much greater, spurting to a final increase of almost 100% over the second mean level and $2\frac{1}{2}\times$ higher than the level at the start of the experiment. A third increase was then planned, and this also led to a rise, this time to a mean firing level of almost 3 times the level at the start.

The results of one such experiment are illustrated graphically in Fig. 7. The initial level was 12/sec, and it was raised progressively to 14, 28 and 33 per second by delivering shocks to the leg on declining discharges, when the mean frequency fell below 10, 15 and 20/sec respectively for a 10 sec period. The higher the frequency "requirement" to

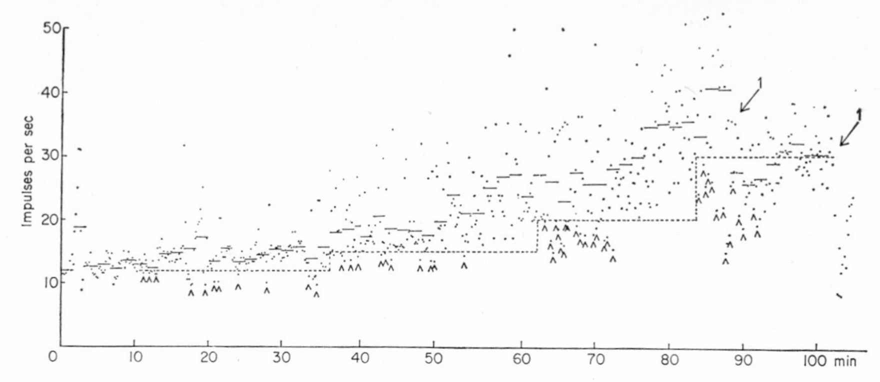

Fig. 7. "Learning". Results of an experiment in which a shock was given to the r. mt. a.c.a. of *S. gregaria* each time its mean frequency over a 10 sec period fell below a prescribed arbitrary level ("demand" level), indicated by the broken line. Moments of stimulation are indicated by arrows. The dots show the mean frequency over 10 sec periods whilst the bars show the mean frequency over 100 sec periods to show the general trend more clearly. After a time the demand level was raised. Note that after a few stimuli have been received the mean frequency rises. In this experiment major inhibitions (indicated by arrows and Number 1) did not occur until the demand level was raised to 30/sec.

avoid a shock, the greater the mean level reached above this value. The preparation does not respond immediately, and will always receive several shocks at the higher requirement before the mean frequency starts to climb to a "safe" level.

The maximum maintained frequency which was reached varied with the animal. Some could not be pushed above 30/sec whilst others readily gave more than 50/sec. In brief bursts the frequency may rise to more than 200/sec, so the demand is well within the capabilities of the nerves concerned.

F. *Effects of Randomly Applied Shocks*

The effect of shocks given at moments not significantly related to the output frequency could most satisfactorily have been tested by the use of simultaneous test (P) and control (R) animals as Horridge did. However, this would have required an inordinately large amount of equipment. Instead, the same timing of shocks used to obtain successful increased firing in one animal were given directly to another, control, animal in which there was a high probability that they would not, by chance, coincide consistently with declining phases of the motor output. These experiments never led to maintained increase in firing rate. When such an animal was later put on a regime of shocks co-ordinated to output it usually did not show a consistent rise such as would have been

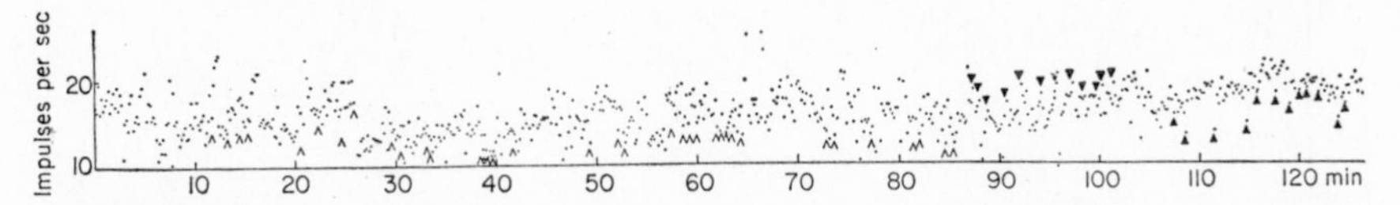

Fig. 8. Stimuli not correlated with output. Stimulus times (indicated by ∧) were taken from a successful experiment of the kind given in Fig. 7. They were thus not correlated with output for the test animal and no maintained increase in frequency resulted. Subsequent stimuli selectively timed to follow rises in mean frequency (indicated by ▼) failed to depress the background and stimuli timed to follow falls in mean frequency (▲) caused only a small rise in the background frequency.

expected had the initial, random shocks not been given. An example is given in Fig. 8. Horridge (1962) had noted that it was more difficult to train R animals and the observations may be related.

G. *Effects of Shocks Timed to Coincide with Rises in Frequency*

Rather variable results were obtained when shocks were timed to occur during peaks in the spontaneous discharge. In a few cases progressive reductions in frequency have occurred, but usually the effect is abrupt. That is, the frequency almost immediately falls to a low level, or to zero, and remains there for some time (Fig. 9). Rarely, the return is abrupt; commonly, the frequency rises slowly back to the background level. Both the duration of the maximum inhibition and the rates of recovery are affected by the number of shocks applied.

In *Melanoplus*, a single shock of adequate strength can, if suitably timed, cause an inhibition to zero which lasts for 2–3 minutes, followed by a slow return to normal background. A pair of similar shocks delivered at a one-second interval causes a slowing of the rate of return by a factor of some 70% (Fig. 10).

H. *Characteristics of the Rise in Mean Frequency*

Although the rise in mean frequency usually shows a delay period, in which the leg receives many shocks, it may occasionally start immediately. In about 30% of the experiments no rise can be obtained. The slopes of the rises show interesting characteristics. They increase at

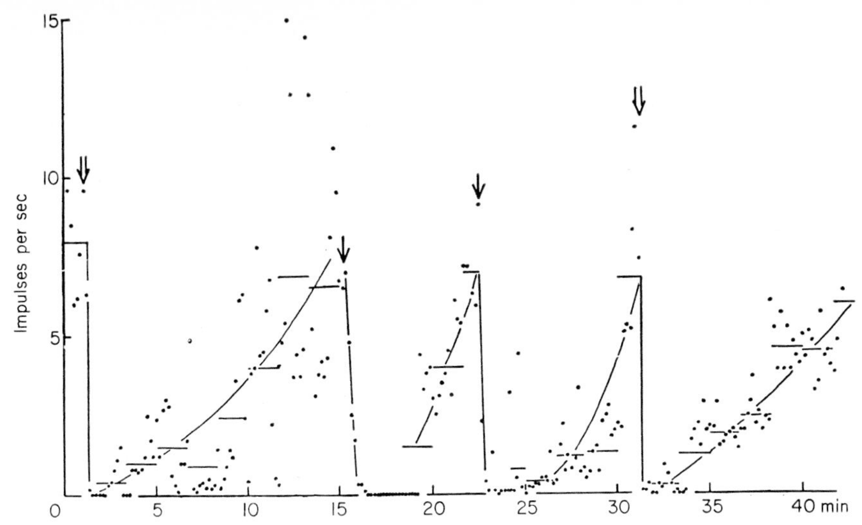

FIG. 9. Central inhibition. Very strong stimuli were applied to the metathoracic leg of *Melanoplus differentialis* whilst recording from the right anterior coxal adductor. The double-stemmed arrow means two shocks were given with a one-second interval. Single-stemmed arrows represent a single shock. Points denote mean frequency over 10 sec. (Bars denote mean frequency over 100 sec.) Note the long time taken to return to background level of 8/sec, and the slow overshoot in the recovery. The recovery time is approximately twice as large after two shocks as after one.

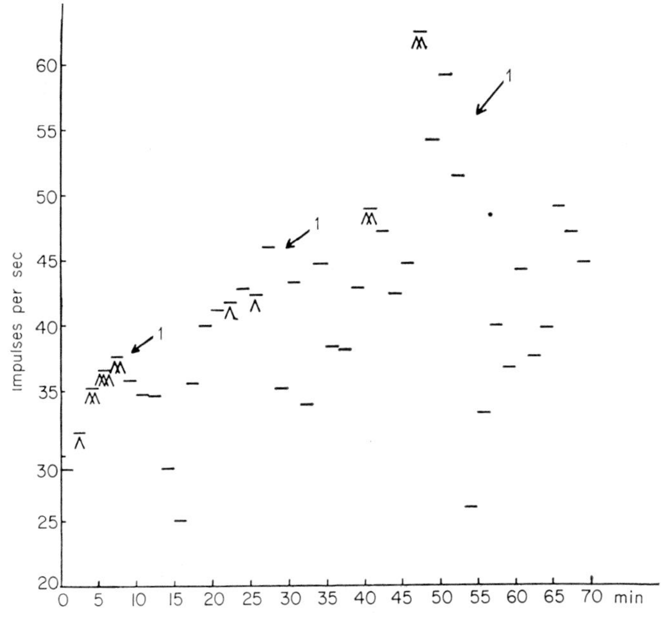

FIG. 10. Temporary inhibition followed by a rise to the level which would have been expected in its absence. In these records (from *S. gregaria* r. mt. a.c.a.) it is evident that inhibition is an independent process which is superimposed upon the units of the nervous system concerned with the learning phenomenon under examination. Bars give mean frequency over 100 sec periods.

first with increasing "demand" frequency—the mean frequency below which shocks are given—but later decline again. The increase takes place gradually, over a period of about 10 minutes, then levels off abruptly.

I. *Limits of "Training"*

A preparation which shows a background discharge of less than 10/sec can usually be made to fire at about 30/sec. Those initially firing at 20/sec or more can be taken to as high as 50/sec, but in most experiments the attempt to push the frequency above about 350/sec leads to marked central inhibitory reactions (Figs. 7 and 13–16). These may interfere for only a few minutes, to be followed by a return to the initial frequency, or even a higher one, such as would have been expected in the absence of such inhibition. However, acute inhibition can last for half an hour or more, and partial depression may last indefinitely.

J. *Duration of Training*

The effect of training in raising the frequency of firing may decline abruptly (probably indicating central inhibition) or slowly, with a time-constant of about $1\frac{1}{2}$ hours. In a few instances a high level has been sustained without significant decline during the whole remaining useful life of the preparation (about 8 hours).

K. *Maximum Rate of "Learning"*

The optimum stimulation to achieve fast learning apparently can only be arrived at by a process of trial and error and the use of intuition. The combination of stimulus strength and duration may be critical. The interval between shocks is also important, but the most important and difficult matter to judge is the precise moment to apply the shock. The total number of impulses given in the learning period should be kept minimal. The strength and duration of the shock are adjusted until a marked burst of impulses is evoked reflexly, without any inhibitory effects accompanying it. If no response occurs, two shocks are tried, at 1 sec intervals, and if two do not cause a response three are given, at 1 sec intervals. When a suitable stimulus combination has been arrived at the preparation is left for a few minutes to recover. Then, when the background rate is steady, the output is monitored carefully. If a frequency decline occurs and lasts for 3 or 4 seconds stimulation is given. No further stimulation is given for a least 3 sec, but at the next tendency to decline another stimulus is given. This procedure is repeated even though the mean frequency may quadruple, during the next 5 minutes, every time a significant fall in frequency occurs. The result of this treatment is a rapid, progressive rise in the rate of firing (Fig. 11).

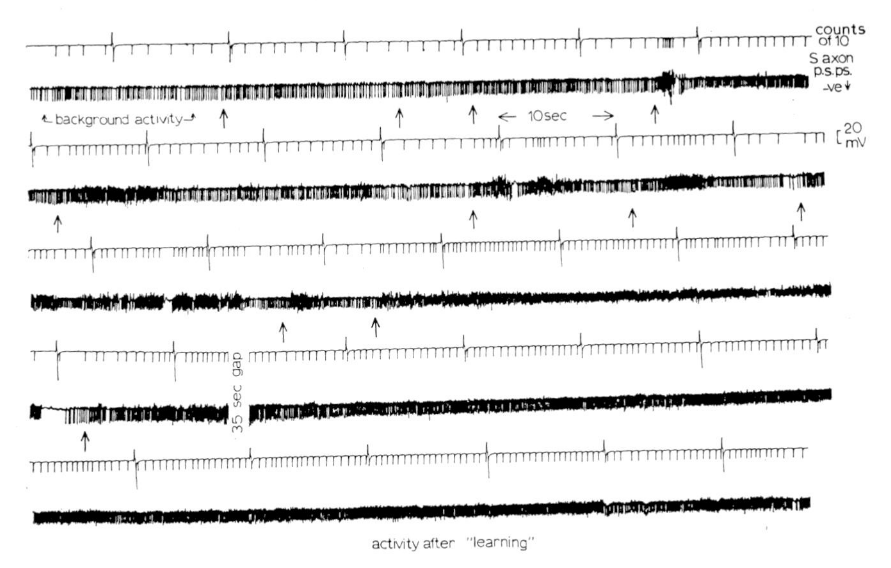

FIG. 11. Maximum rate of "learning". Results of an experiment on the right metathoracic anterior coxal adductor of *S. gregaria* in which shocks (indicated by arrows) were applied to the leg at carefully judged moments, when an otherwise rising frequency started to fall appreciably compared with the previous 5–10 sec period. Electrical signals are intracellularly recorded junctional potentials; depolarization is downward. Background frequency at start was 6/sec. After 11 selectively timed shocks applied over a period of less than 4 min the background was raised to 20/sec. Upper trace: small downward pulses indicate every 10th pulse, larger artifacts coincide with print-out every 10 sec. The smaller potentials of opposite polarity are hyperpolarizing pulses, but they cause mechanical enhancement, not inhibition.

Reducing the interval between shocks, or increasing the total number of shocks received, does not accelerate the change and may impair it. The rate of rise of frequency is steady, when averaged over a 10 sec period, and occurs at the rate of about 20–30% per min for about 4 minutes, thereafter progressively slowing to and reaching zero by the eighth minute. The increase during the second or third minute is usually greatest.

L. *Comparison of Electrical and Natural Stimulus*

Pinching the foot lightly has a comparable effect to electric shocks in leading to an increase in the frequency, and by timing the application of pinches in a similar manner to the timing of shocks it is possible to obtain a similar curve of increasing frequency (Fig. 12). The increase is also maintained.

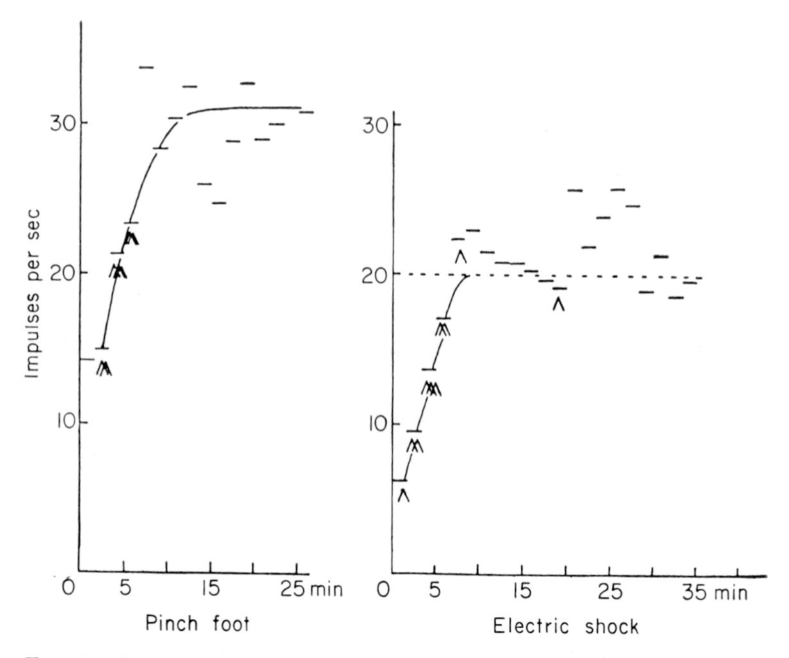

FIG. 12. Comparison of electrical and natural stimulus. Two preparations of right mesothoracic anterior coxal adductor of *S. gregaria*. One specimen (results on left) was given light pinches to the foot at times selected for tendency of frequency to maintained fall. ∧ indicates moment of stimulation. Bars show mean frequency over 100 sec periods. The other specimen (right) had a low background frequency and was given electrical shocks. Broken line shows final mean, which was maintained 2 hours.

M. *Correlation of Discharges in mt. a.c.a. and mt. p.c.a.*

Although the frequency of the S axon discharge in mt. p.c.a. was always higher than that of the single axon supplying mt. a.c.a. the two always changed very closely in parallel. Fluctuations occur approximately synchronously, hence a series of shocks timed appropriately for one is also appropriate for the other. As might be expected, the learning curves are also roughly parallel (Fig. 13). When inhibitions occur spontaneously they also affect each muscle about equally.

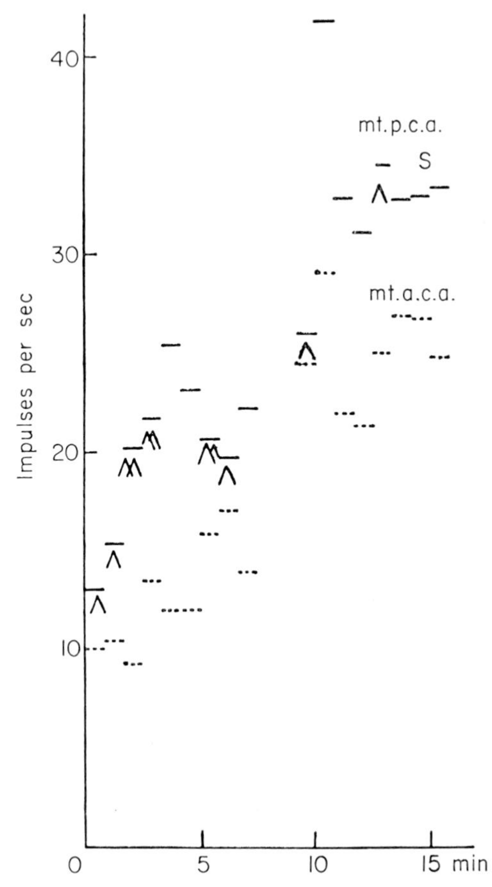

FIG. 13. Correlation of discharges in anterior and posterior coxal adductors. Results from metathoracic leg of *S. gregaria* during fast learning experiment. The frequency followed in the case of the posterior adductor was that of the S axon discharge only. Note that it is higher than that of the single axon supplying the anterior muscle.

N. *Spontaneous and Evoked Central Inhibitory Effects*

It will be seen by glancing at Figs. 7 and 13–16 that inhibitory effects (arrows labelled 1) occurred spontaneously and this was true for all three species. A number of experiments were hampered or ruined by such occurrences. There was, in many cases, a correlation with the stimulus, especially when the rate of firing was already high (above 30/sec), but otherwise the inhibitory effects occurred apparently completely randomly.

Two kinds of effects were recognized, and seem to represent distinct mechanisms. One is a brief effect, reducing the frequency abruptly, but only for a few seconds. The other also occurs abruptly, but declines slowly, taking at least 10 minutes, as judged by the return towards basic mean frequency.

Inhibitory effects have also been observed during studies in which electrodes were placed simultaneously in the homologous muscles of

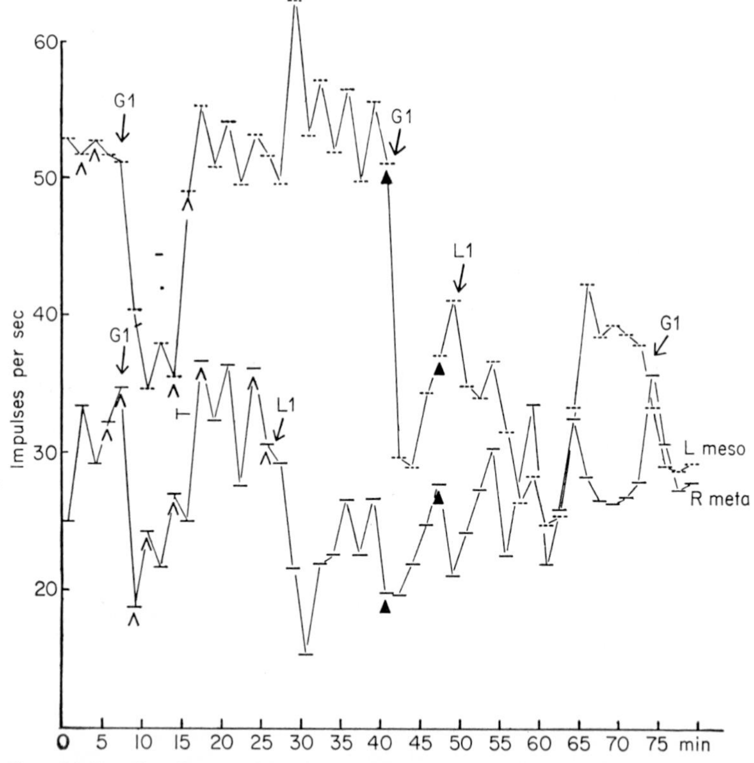

FIG. 14. Local and general inhibitory effects. Mean counts over 100 sec period (bar) from anterior coxal adductors of left meso- (upper at start) and right metathoracic legs. Former had previously been given stimuli so as to raise background level. Single stimuli were applied to either leg as indicated by a ∧ pointing to appropriate trace. Single stimuli were applied to both legs simultaneously at solid ▲. Inhibitions at G1 caused marked fall in frequency to both muscles simultaneously and are termed general inhibitions. Inhibitions at L1 affected the output to only one of the two muscles and are termed local inhibitions.

other legs. These results demonstrated the existence of separate inhibitions which affected either only one muscle, or both simultaneously. The former will be termed local inhibitions; the latter general inhibitions, since they affect at least two muscles on opposite sides and in different segments (Fig. 14). The time-courses of general inhibitory actions affecting output to two muscles are very closely parallel.

O. *Effects of Experiments on Motor Output in Other Joints*

1. *Homologous muscle in same segment; metathoracic legs.* Shocks applied to the leg of one side cause discharges to occur in the coxal adductor muscles of both sides. The increase is not sustained on the non-stimulated side, and falls back to the background level. If, now the contralateral leg is shocked, even randomly, the discharge in the side first stimulated is re-inforced (Fig. 15).

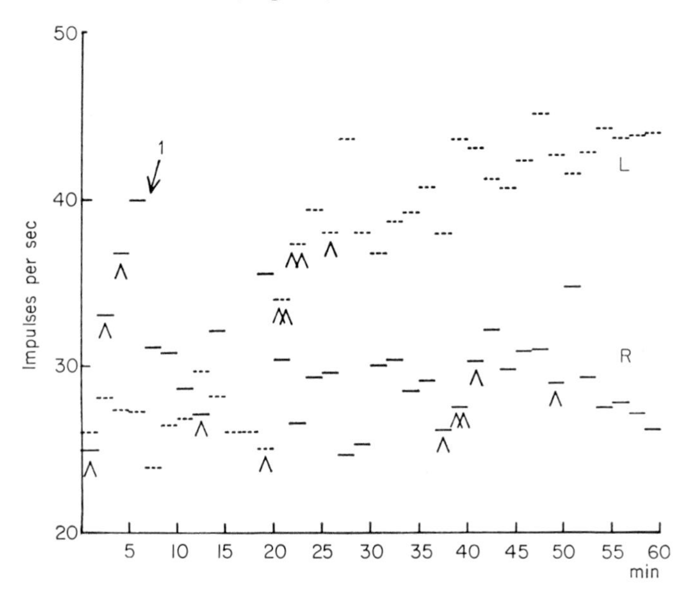

FIG. 15. Correlation of discharges in right and left metathoracic anterior coxal adductors of *S. gregaria*. Bars give mean frequency over 100 sec periods: broken bars left, solid bars right. ∧ denotes moments of stimulation. Note that stimulation was applied first to right leg, then changed to left leg when strong central inhibition occurred.

The contralateral side can be trained to fire faster just as if it had been unaffected by training stimuli applied to the other leg. The background discharges of the muscles on the two sides may be markedly different at the start, for instance as different as 8/sec on one side and

28/sec on the other. By appropriate timing of shocks they can be made equal or even reversed.

2. *Homologous muscles on same side of animal.* The discharges in the mesothoracic coxal adductors are not closely similar in different legs on the same side. Nevertheless, a shock applied to one affects the other, usually causing a brief increase in discharge rate only.

3. *Muscles of different segments on opposite sides of the animal.* Several studies have been made on the effect of shocks applied to one mesothoracic leg on the discharge in the metathoracic leg of the contralateral side and vice versa in locusts. The reason for these experiments was that Horridge (1962) found that a re-test using a different leg of a trained animal showed that it learned quickly, indicating a transfer of some kind. In the present experiments, there was no significant long-

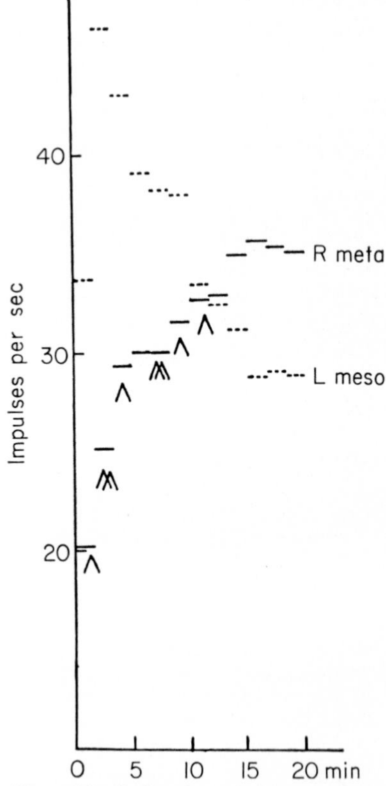

FIG. 16. Influence of fast learning experiment in r. mt. a.c.a. on discharge to l. ms. a.c.a. of *S. gregaria*. The shocks led only to a temporary increase in discharge rate in the latter.

lasting effect in either direction. This suggests a relative independence of the hind leg of locusts, which is hardly surprising in view of their specialized nature and jumping function. No attempts have yet been made to do the experiments with cockroaches. However, short-lasting effects were commonly encountered (Fig. 16).

P. *Learning in the Absence of Proprioception*

A feature of Horridge's experiments was a probable association of the shock with leg position as such and therefore a dependence on afferent information. With the exception of a few abdominal muscles (Osborne and Finlayson, 1958) it is not considered that insect muscles contain stretch receptors. Instead, companiform sensilla in the joints and hair plates subserve the proprioceptive function, possibly aided by some modified chordotonal organs.

The present experiments were done with legs fixed so that leg position could not have been significant in determining the results. However, a change in the frequency of discharge must be reflected in a change in discharge from the campaniform sensilla, which are variably stressed depending on contraction of the muscle even under isometric conditions. Changes in the frequency of this discharge could still be of importance in relation to the learning process.

The effect of sensilla associated with the distal attachments of the adductors could be eliminated by cutting the muscles free at their distal apodemes and attaching the cut apodeme of a test muscle to a microclamp. Preparations of this kind were made with the mt. a.c.a. and the fast learning procedures applied (Fig. 17). The best curves obtained were not significantly different from the curves obtained with the

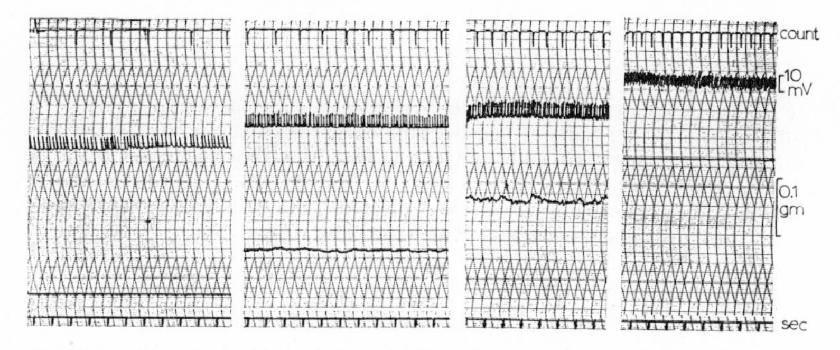

FIG. 17. Learning in absence of proprioception. Right anterior mesothoracic coxal adductor of *S. gregaria*. Apodeme was severed and clamped by force transducer. Records show samples taken at different stages during rate increases associated with training. Tension at left is almost zero, frequency 5/sec. Tension at right has increased to 0·2 gm; frequency is 20/sec.

apodeme intact. It is concluded that position effect is probably not significant in determining the results. Further work is needed to test the role, if any, of proprioception in Horridge's experiments.

Q. *A Comparison with the Results of Horridge*

When the results of the present experiments on fast learning in mt. a.c.a. were averaged and a smooth curve drawn between the points a simple exponential curve was obtained (Fig. 18). The frequency rises to its

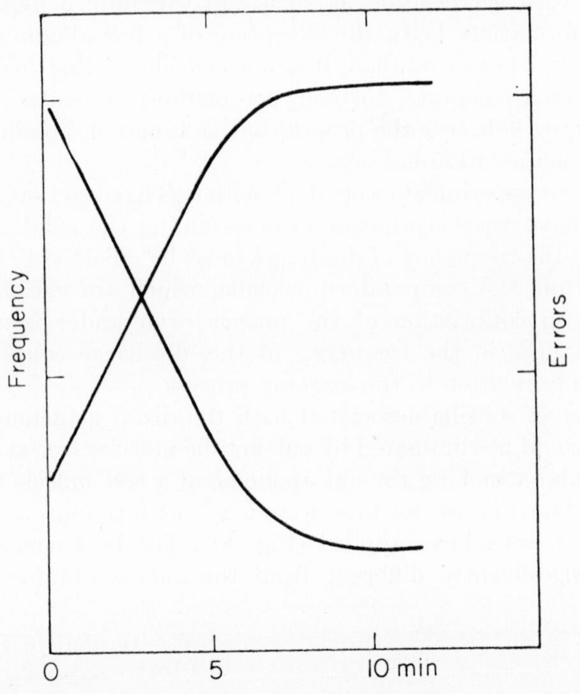

FIG. 18. A comparison of electrophysiological data curves with results of Horridge. The results of fast-learning curves from the present experiments were smoothed and plotted against time (rising line). On the same graph a smoothed curve based on Horridge's experiments of the number of errors per unit time is also plotted (falling line). Note that the curves are nearly reciprocally related.

maximum in about 8 minutes if stimulation is stopped after 5 minutes. It is interesting to find that if the published results of experiments on the locust mesothoracic leg are smoothed and plotted on the same graph, the curves of errors per minute against time and frequency per second against time are reciprocal. In particular it is interesting that both curves flatten after 8 minutes.

Leg extension after the initial shocks in Horridge's experiments could be the result of random increases in the frequency of discharge of extensor muscles. The ability of these discharges to cause leg extension will be reduced as the frequency of firing in the flexors rises. Also, there may be a reciprocal inhibitory connection between the flexor and extensor motorneurones, leading to a reduced tendency for firing in the extensors. Only the highest frequency extensor bursts will overcome flexor resistance as the latter increases. Thus we could predict a progressive reduction in the number of extensions (leading to "errors") in the type of experiments done by Horridge on the basis of the mechanism discovered in the course of the present work and are in a position to suggest tentatively that it is the cause of the "learning" phenomenon.

R. *Spontaneous Central Excitation*

Since general inhibition of the background discharges has frequently been observed one would expect increases in excitatory level perhaps to be equally common. In fact, these are rather rare, which is fortunate for the success of the experiments. They are much more frequent, as are all fluctuations, when the head is not removed, presumably in relation to the animal's attempts to escape. Immediately following decapitation there is always a greatly increased discharge rate, and this may take over half an hour to subside (Fig. 19). This experience means that within the thoracic ganglia there are neurones which are capable, when excited as a result of decapitation (perhaps a release of inhibition?), of raising the firing level over a long period, albeit at a declining frequency. When the connectives between the metathoracic ganglion and the mesothoracic and abdominal ganglia are cut, thereby isolating the ganglion in which the relevant motorneurone is located, the discharge remains unchanged for a few minutes. Thereafter it declines rather rapidly to zero. When a metathoracic ganglion of a preparation trained to fire at a high frequency was suddenly isolated, and all nerves cut except the one containing the relevant motorneurone, there was no immediate change in frequency. It is thus possible to conclude that the pace-setting of the coxal adductor excitatory neurones is determined from within the same ganglion which contains the neurones.

V. Discussion

The present work has revealed a process in insects which helps to explain a phenomenon which has long been known: namely extreme plasticity in the control of appendages. It is evident that this is due

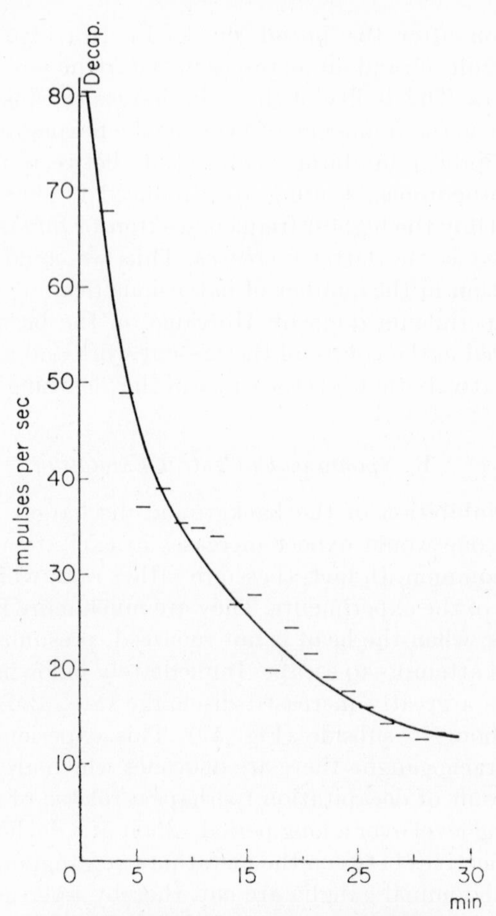

FIG. 19. High rate and subsequent decline in frequency following decapitation. Bars give mean frequency of discharge in right anterior mesothoracic coxal adductor of *S. gregaria*/100 sec. Initial background frequency of 12/sec was not recovered until 30 min following decapitation.

to a large degree of segmental independence so that even after the removal of high-level overall control mechanisms, presumably located in the brain, beautifully co-ordinated movements can still occur. Furthermore, even within the nervous system of a single segment long-lasting changes of adaptive significance can occur in response to external events. These are equally readily reversed should the relationship with the environment change.

These changes require an afferent nervous discharge such as is caused by a sharp pinch or externally applied electric shock to follow closely upon an efferent one. The afferent channel need not be a specific one: an electric shock given to the whole leg can hardly be specific in eliciting discharges. Certainly it acts on nerves which are below the level of proprioceptors which might be useful in indicating leg position. Horridge (1962) supposed that leg position, as indicated by sensory discharges from coxal hair plate cells, might be an essential feature of learning to maintain a leg in the raised position by headless insects poised over a bath carrying the means of delivering electric shocks following leg extension. This might be true, for experiments on the whole leg, but the present experiments eliminated the leg position aspect and showed that a long-term learning phenomenon can nevertheless take place. Furthermore, a similar phenomenon could, in principle, account for the results of Horridge's experiments.

The essential feature in both the plasticity and the learning phenomena could be a common aspect of neuropile function. That is, as a result of possible widespread interconnections between neurones there will be some influence of any afferent input on any efferent output in insect nervous systems. Evidence for such widespread interconnections has been given (Hoyle, 1964). It has been shown in the present work that long-term effects do not usually occur if strong shocks are given (strong afferent discharge from many receptors of a limb) at random intervals. The only effect these may have is a temporary excitation or a moderately prolonged inhibition. The long-lasting changes occur only when shocks consistently follow a significant trend in the motor output. This trend may be an increase or a decrease, though the present experiments have been mainly concerned with the effects of stimuli following decreases.

Whatever the short-term effect of the timed shocks the long-term one has always been a trend towards the converse of the situation which led to the shock being received. That this phenomenon is adaptive is self-evident. There is also no doubt that it is a form of learning. There has been no general acceptance of the phenomenon in Horridge's experiments as an example of learning, and indeed this was anticipated by Horridge in his discussion. In part this is due to there being several different kinds of phenomenon to which the term learning may be applied. The phenomenon Horridge discovered had about it the aura of a simple physiological mechanism, perhaps a peculiar feature of insect nervous systems and not of general significance. The present work has emphasized these possibilities, though in doing so perhaps the almost mystical quality which has come to surround the unknown learning process may have been lessened. Many examples of elementary learning processes in higher animals are of a similar nature in principle and the insect example could serve as a model for them. They include learned avoidance reactions, some postural activities and orientations and perhaps some forms of goal-seeking.

The importance to general physiology of this insect phenomenon is that it provides the promise of a preparation which will permit a study of the cellular changes involved in a long-term change in the nervous system. All forms of learning must involve long-term changes, and although it is not hard to conceive of ways in which they could conceivably occur (modern models range from reverberatory neuronal circuits to changes in neuronal nucleic acid structure) no actual examples are known. The present work has narrowed down a phenomenon of long-term central neuronal change to a situation in which it is fairly readily reproducible in material available to many workers and in which the final indicator is a single cell whose effects are easily tapped.

Working from this point on may be more difficult. The motorneurone itself can be located by simply exciting anti-dromically from the level of the muscle but may not be the site of the underlying changes and interneurones concerned in the process may be impossible to locate. Another possibility is that synaptic elements in the neuropile are concerned, and no method has yet been devised for studying these effectively. Further explorations are being carried out to try to locate sites in the neuropile where electrical activity correlated with the output of the excitor neurone to the metathoracic anterior coxal adductor of *Schistocerca* may be recorded.

VI. Summary

1. The mechanism of "learning" to lift a leg in headless insects (the phenomenon found by Horridge) has been examined by electrical recording from various muscles which might be utilized. The coxal adductors show a maintained discharge which has a high frequency after training and so may play a significant role in the process.

2. A preparation is described whereby the excitatory neural discharge to the coxal adductors is monitored by intracellular recording. The discharge is due to one or two neurones only and can be electronically counted.

3. A spontaneous background discharge is always present in coxal adductors of *Schistocera gregaria* and *Melanoplus differentialis*. If single supra-threshold electric shocks are given to the tibia/tarsus, brief central excitatory or inhibitory effects occur and are reflected in the discharge.

4. When shocks are timed to follow significant spontaneous falls in the

frequency of the background discharge a rise in frequency results and the frequency rise may be maintained from several seconds up to minutes or rarely hours. Further selectively timed stimuli lead to further rises and longer maintenance.

5. The process of giving selectively timed shocks can be repeated until the frequency has been raised by as much as 3 or 4 times, from 5–20/sec at the start to 30–60/sec at the limit of training.

6. The higher frequency background discharges are interrupted from time to time by spontaneous central inhibitions. Central inhibitions may also be induced by giving very strong shocks to the tibia/tarsus.

7. The converse training, namely "learning" to produce a lower frequency, may be caused by timing shocks to arrive following spontaneous bursts at higher than average frequency. Then the output is depressed and a trend towards maintained high frequency can be reversed.

8. Stimuli applied to one leg affect the motor output on the ipsilateral and also the contralateral side, but not in any simple, consistent way.

9. The shortest time in which the frequency can be raised to the maximum to which it can be driven in a given case by applying selectively timed shocks to a leg is limited to about 4 minutes, except in rare cases where a single shock leads to a prolonged response, and usually takes 8–10 min. The average time to reach maximum is 8 minutes.

10. It is proposed that the phenomenon of frequency increase following timed shocks can account for the learning phenomenon in headless insects discovered by Horridge. The time-courses of the two are similar.

11. It is further suggested that this will prove to be a general phenomenon, at least in insects, and to have adaptive significance in several contexts.

ACKNOWLEDGMENTS

Locusts were very kindly supplied by the Anti-locust Research Centre, London, by air mail. *Melanoplus* were descended from a stock originally provided by Dr Eleanor Slifer.

REFERENCES

Campbell, J. I. (1961). The anatomy of the nervous system of the mesothorax of *Locusta migratoria migratorioides* R. & F. *Proc. Zool. Soc. Lond.* **137**, 403–632.

Cohen, M. J., and Eisenstein, E. (1965). Learning of leg position by single segments of cockroach. (In preparation.)

Horridge, G. A. (1962). Learning leg position by the ventral nerve cord in headless insects. *Proc. roy. Soc.* B. **157**, 33–52.

Hoyle, G. (1953). Potassium ions and insect nerve muscle. *J. exp. Biol.* **30**, 121–135.

Hoyle, G. (1957). The nervous control of insect muscle. *In* "Recent Advances in Invertebrate Physiology" (B. T. Scheer, ed.), pp. 73–98. University of Oregon Press.

Hoyle, G. (1964). Exploration of neuronal mechanisms underlying behavior in insects. *In* "Neural Theory and Modeling" (R. F. Reiss, ed.), pp. 346–376. Stanford University Press.

Hoyle, G., and Smyth, T. (1963). Neuromuscular physiology of giant muscle fibers of a Barnacle, *Balanus nubilus* Darwin. *Comp. biochem. Physiol.* **10**, 291–314.

Osborne, M. P., and Finlayson, L. H. (1962). The structure and topography of stretch receptors in representatives of seven orders of insects. *Quart. J. micr. Sci.* **103**, 227–242.

Brain Controlled Behaviour in Orthopterans*

F. HUBER

Institute of Comparative Physiology, University of Cologne, Germany

The brain of insects is known to be an important link in the control of behavioural activities by descending inhibitory and excitatory commands to the ventral nerve cord which organizes the patterns. This control can be widespread or limited depending upon the structural and functional arrangements which the descending fibres make within the ganglia.

In some activities the brain seems to act as a switch by turning on and off the output of motor centres in the cord, or by changing their threshold to input signals of other sources. In other activities the brain determines both duration and strength, and the temporal and spatial pattern of behaviour. This paper deals with mechanisms in the CNS of crickets (*Gryllus campestris* L.), in which the brain is involved in the inhibitory control of evasive behaviour and in the regulation of sound production in the male.

I. Evasive Behaviour

Roaches, crickets and some other groups of insects have developed an ascending giant fibre system within the abdominal nerve cord known to be an essential link in the neural mechanism of escape behaviour. Within the last abdominal ganglion the giant internuncials receive input signals from sensory fibres of the anal cerci, and within the metathoracic ganglion the giant fibres are connected with motorneurones supplying the leg muscles. Roeder (1948, 1959) has studied the way in which signals are carried and transmitted from the cercal afferents *via* the giant fibres to the motor efferent fibres in the cockroach. In crickets we have started a detailed electrophysiological analysis.

An undisturbed male or female cricket responds to cercal stimulation with kicking of the hind legs. Strong mechanical stimuli evoke movements in both legs which are often followed by a forward locomotion.

* This investigation was supported by a grant from Die Deutsche Forschungsgemeinschaft.

Evasive behaviour can also be elicited through mechanical stimulation of the antennae, but it differs from that due to cercal stimulation in the absence of kicking. The kicking response can be elicited in a 1:1 relation, if the stimuli applied follow each other in intervals longer than 30 seconds. Touching the cerci in shorter intervals gives rise to

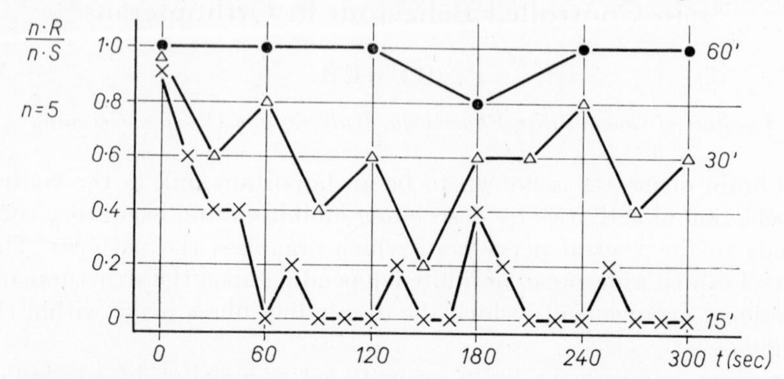

FIG. 1. Kicking responses as a function of mechanical stimulation of the cerci at different intervals. Each point on the curves of this and the following figures indicates the number of responses (n.R.) per number of stimuli (n.S.) applied. The number of stimuli is shown to the left of the ordinate.

a fatigue which is apparently due to changes in the properties of the synapses involved (Fig. 1).

The cercal response in male crickets varies in threshold and may even disappear during certain sexual activities (Fig. 2). An adult male

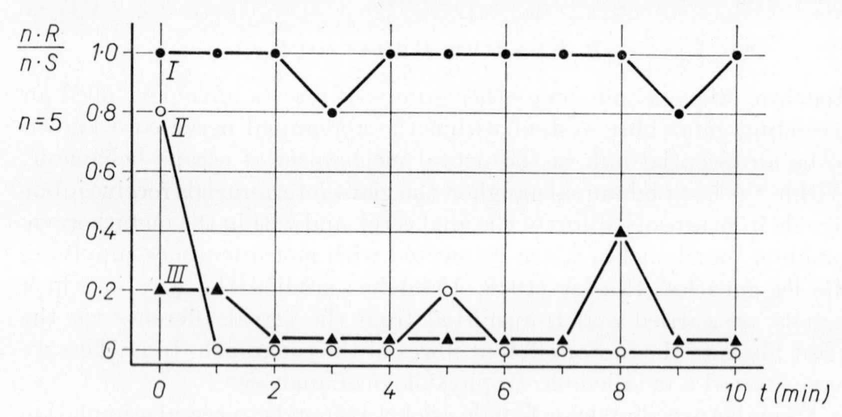

FIG. 2. Kicking responses as a function of different states of sexual responsiveness. Mechanical stimuli follow in intervals of one minute.

cricket goes through different stages of sexual responsiveness. The "unresponsive state" is marked by the absence of stridulation, although a spermatophore is present, and kicking can easily be elicited. The "responsive state" is indicated by calling and courtship songs which precede copulation. Cercal stimulation evokes copulatory movements and suppresses singing and kicking. During post copulatory activity (e.g. in the absence of a spermatophore), the male responds either with a specific turning manœuvre towards the female or more often with fighting songs (Fig. 3). These observations lead to the conclusion that

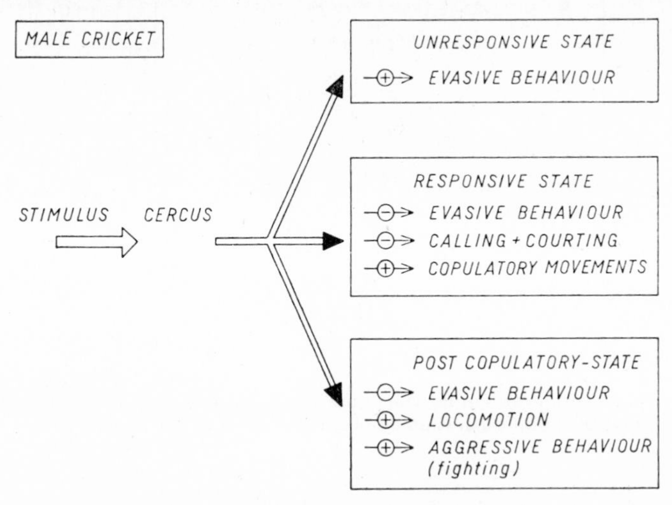

FIG. 3. Types of reactions due to cercal stimulation in male crickets during certain states of sexual responsiveness.

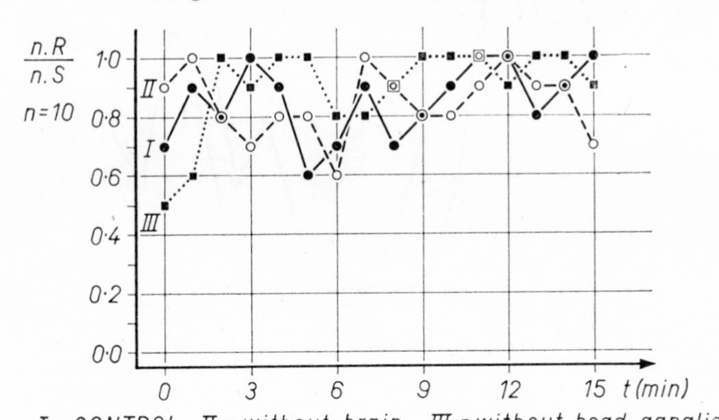

FIG. 4. Kicking responses before (I) and after (II) ablation of the brain and the suboesophageal ganglion (III). $P_{(I-II)} = 0{\cdot}81$; $P_{(I-III)} = 0{\cdot}48$; $P_{(II-III)} = 0{\cdot}74$.

kicking disappears as soon as the male changes from the unresponsive to the responsive state. We have investigated the neural mechanism underlying this inhibitory process.

After removal of the brain and the subœsophageal ganglion, crickets show a slight increase in the kicking response as compared with the controls (Fig. 4), which is not statistically significant (P = 0·48). Fatigue was, however, less often seen when stimuli applied to the

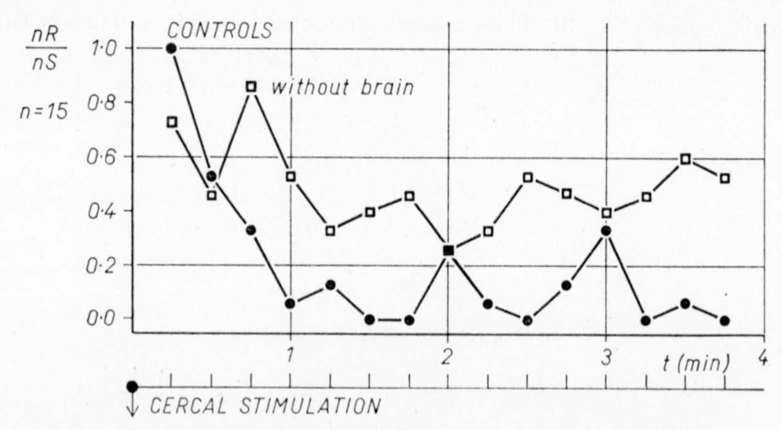

Fig. 5. Different rates of fatigue in the kicking response of males with and without the brain when stimulated at intervals of 15 seconds. P = 0·01.

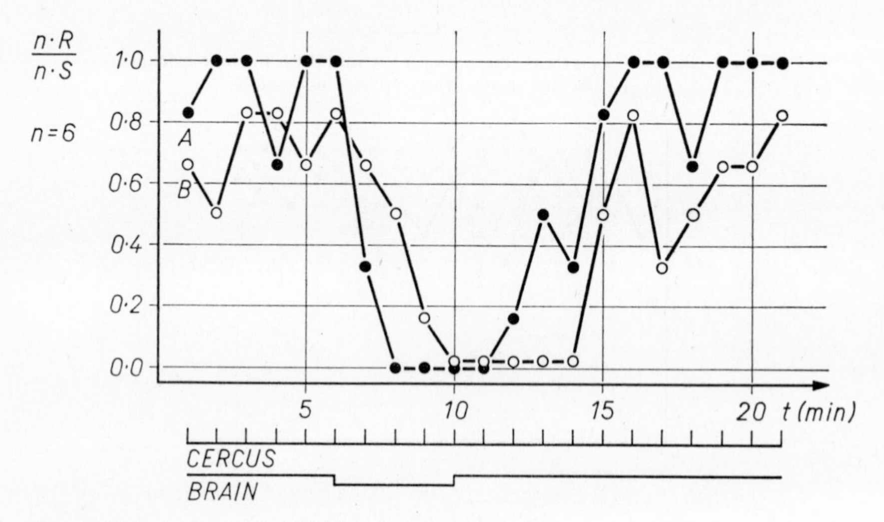

Fig. 6. Suppression of the kicking response during electrical stimulation within the calyx of the mushroom body. Mechanical stimuli follow at intervals of 1/min. Brain stimulation: square pulses, 0·1 ms, frequency 50/sec, current strength 9 μAmp.

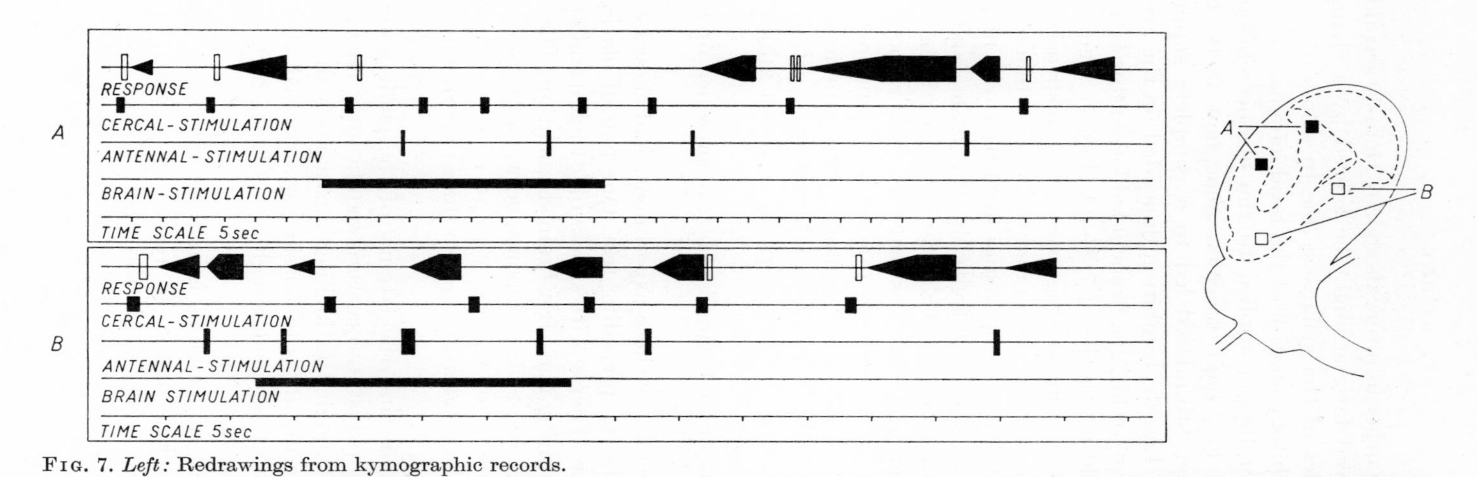

FIG. 7. *Left:* Redrawings from kymographic records.

(A) demonstrates the absence of both the kicking response □ and locomotor activity ◄ during brain stimulation within the mushroom body (■ right figure);

(B) shows an example of a selective inhibition of the kicking response due to stimulation at other regions (□ right figure). In this case forward locomotion could be elicited through mechanical stimulation of the antennae.

Right: Diagram of the lateral part of the cricket brain with the loci situated within the mushroom body (dotted line).

cercus succeeded each other in intervals shorter than 30 seconds (Fig. 5; P = 0·01). This reduced fatigue might be caused by a change in the transmitting properties of the synapses, due to a loss of a weak but steady operating inhibitory control of the head ganglia.

Focal electrical stimulation applied to the protocerebrum of male crickets kept on a holder have revealed some new facts. Loci were discovered which when stimulated led to a complete suppression of the kicking response (Fig. 6). Inhibition depended on the area excited and was either selective for kicking or included locomotor activity when stimulated at threshold intensities of 7–15 μA (Fig. 7). Neurones which seem to be involved in suppressing the cercal responses could be localized within the glomerular region and the lobe system of the mushroom bodies. As we shall see these bodies form part of a neural mechanism which regulates sexual activity in the male, and to which simultaneous inhibition of escape behaviour apparently belongs.

Roeder (1948) has shown in roaches that the cercal response depends on transmission of signals at the cercal-giant fibre junction in the last abdominal ganglion and at the giant-motorneurone junction situated in the third thoracic ganglion. An essentially similar system seems to exist in crickets. In order to find the place in which inhibitory commands affect the neural mechanism, we first studied the transmission process in the last abdominal ganglion in the presence or absence of brain stimulation.

The four males selected for these experiments had shown a complete inhibition of kicking in the preceding test. One pair of platinum wires served to excite the cercal nerve with shocks of 0·1 msec and at intervals of 30 seconds before, during and after brain stimulation. The other pair was used to record the giant fibre response between abdominal ganglia 2 and 3. The movements of the leg stumps were observed under a microscope. Some of the results obtained in these experiments are shown in Fig. 8A–C.

Neither the amplitude and the latency of the giant potentials nor the number of spikes per stimulus were significantly changed during brain stimulation, although the kicking movements disappeared. In another series of experiments transmission was studied after removal of the brain and subœsophageal ganglion, and again, no significant change could be found (Fig. 8D–F).

These results suggest that kicking is inhibited at the second synapse within the metathoracic ganglion. In roaches this junction has been thought to be the most labile link in the whole system (Roeder, 1959).

Previous recordings from some of the lateral nerves 4–6, which are known to supply muscles involved in kicking, had not given a clear

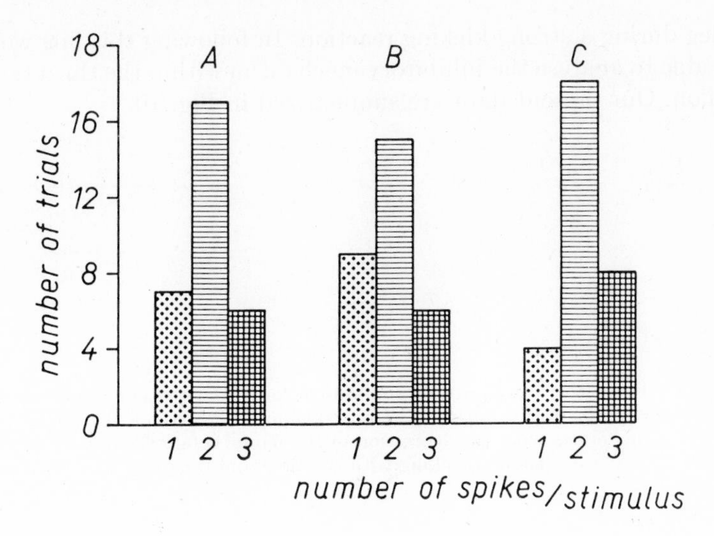

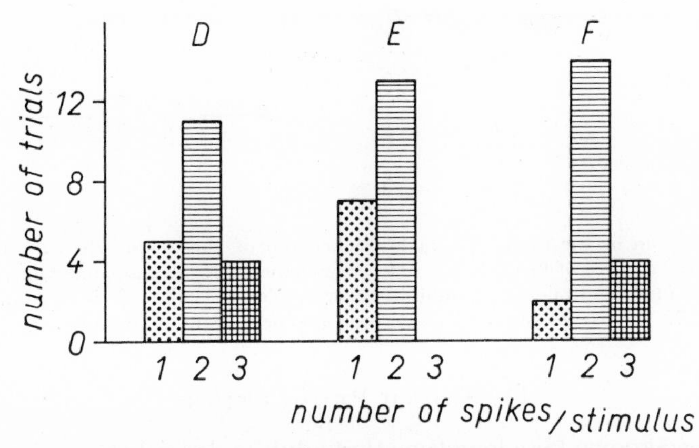

FIG. 8. *Upper diagram:* Histograms of giant fibre responses elicited before (A), during (B) and after (C) brain stimulation through electrical shocks applied to the right cercal nerve at frequencies of 2/min.

Lower diagram: Giant fibre responses before (D) and after (E) removal of the brain and the suboesophageal ganglion (F.)

answer to our problem. We therefore made recordings from muscles in unrestrained crickets. Some of these muscles are innervated by only one or two motor fibres and appear to respond in a 1 : 1 fashion to the arrival of nerve impulses. Figure 9 illustrates the repetitive discharge of one unit belonging to the depressor trochanteris of the left hindleg during a strong kicking reaction. In following this line we hope to be able to analyse the inhibitory mechanism within the third thoracic ganglion. Our present data are summarized in Fig. 10.

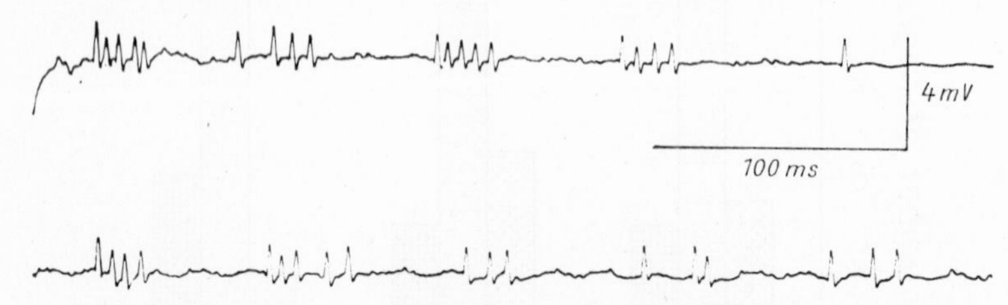

FIG. 9. Repetitive discharges of one muscle unit from the left metathoracic depressor trochanteris during a strong kicking response of the ipsilateral leg. Kicking starts with the depression of the femur before the animal begins sudden kicking with the tibia and tarsus.

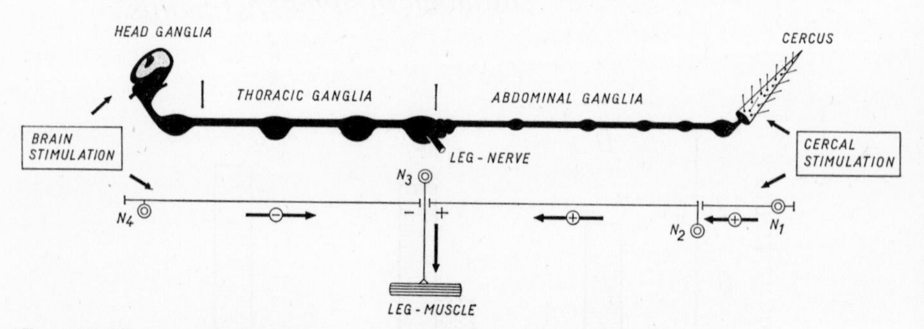

FIG. 10. Diagram of the CNS of a male cricket and of the neural elements or systems involved in the control of kicking. N_1 = sensory element; N_2 = giant internuncial; N_3 = metathoracic motorneurone; N_4 = inhibitory element of the brain. It is unknown whether N_4 represents one neurone on each side or a chain of neurones.

II. SOUND PRODUCTION

It has been known for a long time that adult male crickets produce three types of songs, each of which is characterized by a specific temporal pattern of pulses and trills. These songs serve the purpose of communication between members and sexes of the species (Fig. 11). Calling and courtship songs only start in the presence of a spermatophore. The physiological condition associated with this state seems to be signalled via the ventral cord to the brain. The absence of a spermatophore, as well as severing the abdominal cord, changes the behaviour of the male. Calling and courtship disappear and are followed by the post copulatory activity (Huber, 1955). Both courtship and aggressive sounds are normally elicited as soon as males and females touch each other with the antennae. Whether courtship or fighting occurs depends on the strength of antennal contact and on the kind of response elicited in the touched animal.

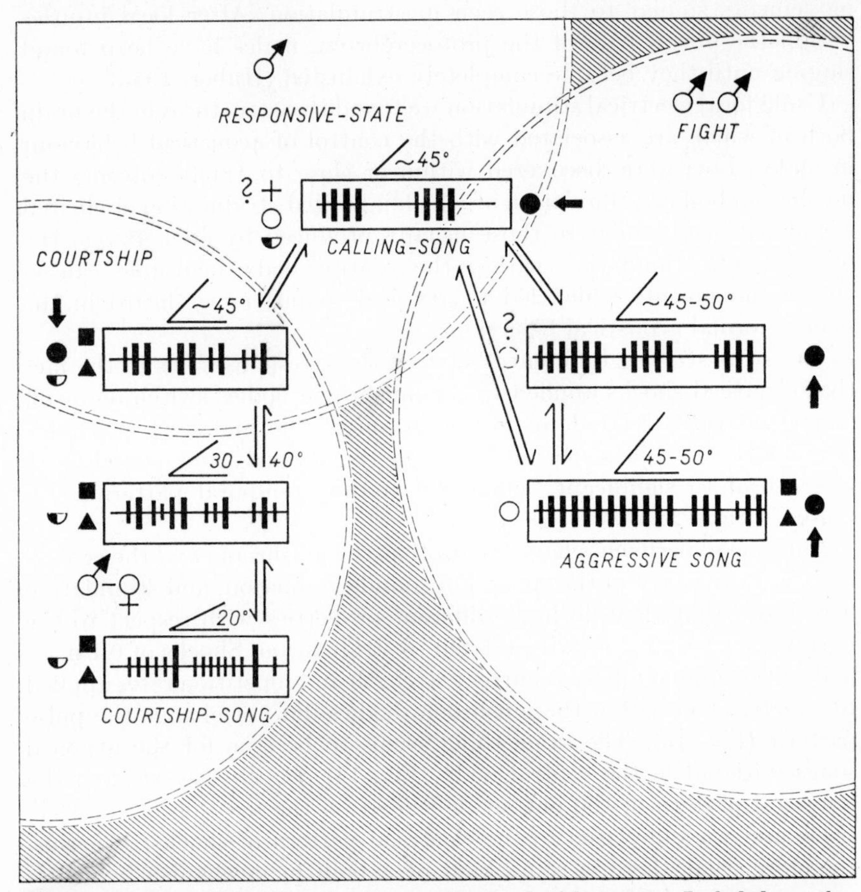

FIG. 11. Diagram illustrating the three patterns of cricket songs. Included are the transitional states between calling and courtship (redrawings from oscillograms). < angle between the body and the raised wings; ◒ songs to be elicited only in the presence of a spermatophore; ■ songs in response to antennal and ▲ to cercal stimulation; ○ songs in response to auditory stimuli; + possible acoustical feedback; ●▲ song patterns elicited by stimulation within areas of the mushroom bodies.

Ablation experiments have shown that male crickets are still able to stridulate in a normal manner with a neural system containing the brain, together with one connective to the second thoracic ganglion and its lateral nerves which innervate the stridulatory muscles. However, if parts of the two mushroom bodies are removed, or if the central body neuropile is destroyed by use of HF-coagulation, singing fails com pletely. The thoracic cord cannot initiate and maintain normal stridulation although strong mechanical stimuli applied to the elytra evoke movements similar to those seen in stridulation. After local injuries within the dorsal part of the protocerebrum, males have been found singing until they became completely exhausted (Huber, 1955).

Using focal electrical stimulation we found two structures in the brain both of which are associated with the control of acoustical behaviour in males. Loci were discovered within or close to tracts entering the mushroom bodies, stimulation of which inhibited stridulation or elicited sound patterns similar to those usually produced by crickets. On the other hand, stimulation within the central body neuropile caused elytral movements which led to atypical sounds never heard in the male's normal acoustical behaviour.

To us the striking difference between these responses lies in the fact that electrical shocks applied to the mushroom bodies switch on or off neural activities related to the formation of "normal song rhythms" (Fig. 12: 1–4), while shocks given to the central body and caudal brain tracts lead to significant "changes" in the "temporal patterning of pulses" (Fig. 12: 5–7).

Ablation experiments have shown that the mushroom and the central body are necessary in the control of sound production, and stimulation has shown that they do have different properties with respect to the patterning of neural activity related to stridulation. Shocks of 0·1 msec in duration and of different current strength and repetition rates applied to effective loci within the mushroom bodies did not change the pulse pattern (Fig. 13). The same thing seems to be true for the atypical songs initiated in the central body (Fig. 14). In all cases studied the temporal pattern of the output depended on the location of the electrode and was unrelated to the electrical pattern of shocks. It seems, therefore, that there is some kind of a built-in mechanism which is responsible for the formation of pulses and trills in calling, courting and fighting. The fact that we were able to change the temporal pattern in the central body neuropile indicates that this region of the brain is closely related to this system. Inhibitory and excitatory control of calling always lasted longer than stimulation in these experiments and was often succeeded by post-inhibitory or post-excitatory rebound activity (Fig. 15b, c). Repeated shocks led to a remarkable decrease in the threshold intensity, and after-effects were noticeable for more than 30 minutes. This indicates a high degree of facilitation within the brain, and the activated system can be compared with an oscillator which is less damped.

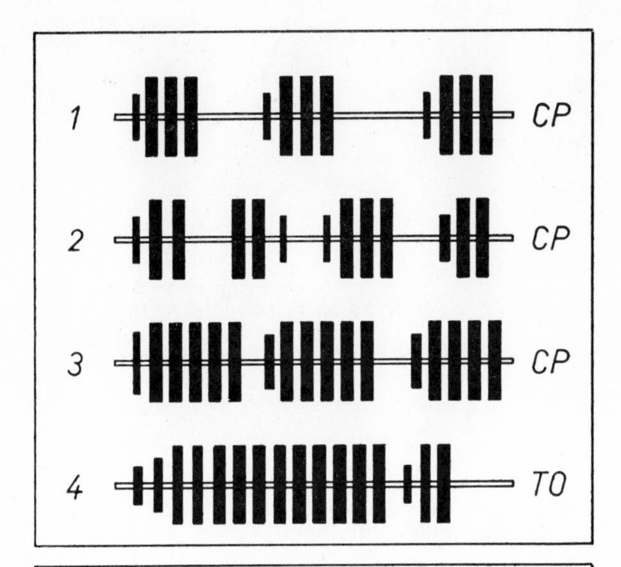

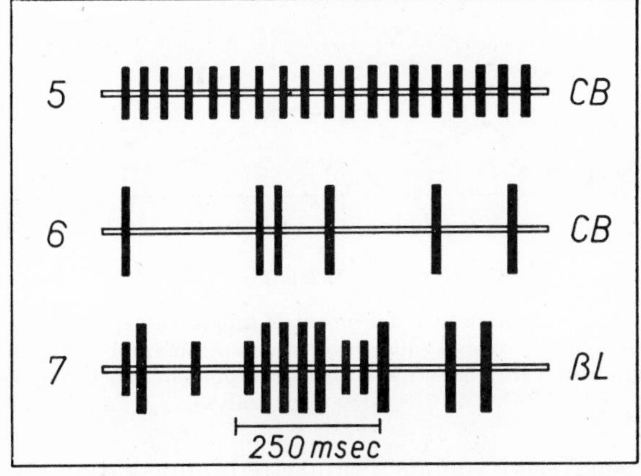

FIG. 12. Redrawings from parts of oscillograms showing song patterns elicited during stimulation within different parts of the protocerebrum. CP mushroom body, TO *tractus olfactorio globularis*, CB central body, β L lower part of the stalk system, 1 calling, 2 the first transitional state from calling to courtship, 3 and 4 aggressive sounds, 5–7 atypical sounds.

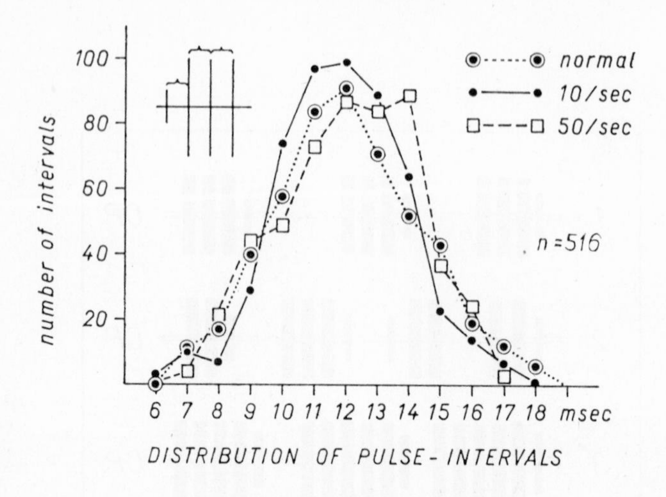

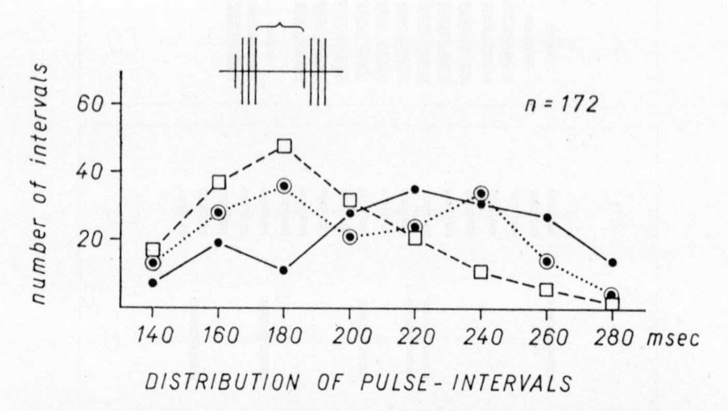

Fig. 13. Distribution of pulse intervals (upper curves) and of intervals between successive trills (lower curves) in normal calling and in calling elicited through stimulation with different shock frequencies at the same locus (calyx).

However, aggressive sounds disappeared immediately after cessation of stimulation, although a tendency for fighting remained several minutes (Fig. 15d, e). The absence of after-effects leads to the conclusion that this mechanism operates only in the presence of input signals, and this hypothesis is strongly supported by observations of the normal aggressive behaviour in male crickets.

Some of the atypical songs were produced by a continuous vibration of the elytra (Fig. 15f). This pattern was changed during the post-stimulatory phase. The male interrupted stridulation more and more

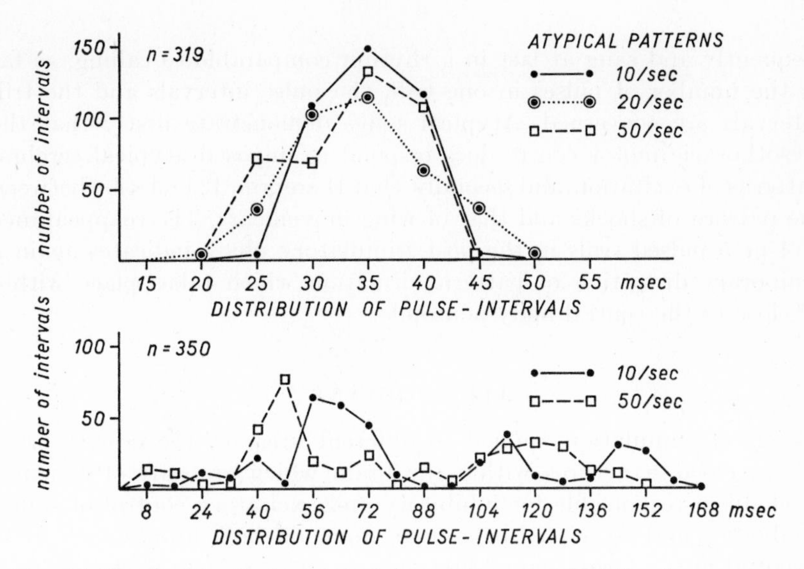

FIG. 14. Distribution of pulse intervals in a continuous stridulation (upper curves) and in an irregular song (lower curves). Electrical shocks of different frequencies have been applied to two closely related loci within the central body.

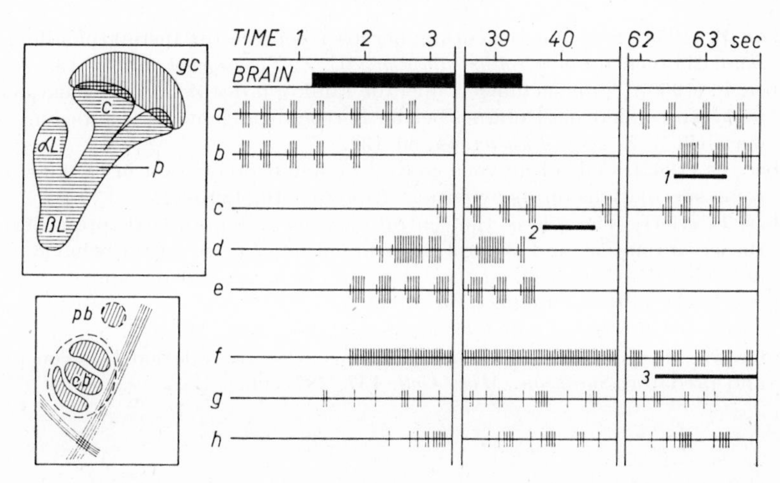

FIG. 15. *Left:* Sagittal view of the mushroom body (above) and the central body with the protocerebral bridge (below). gc = region of the glomerular cells, c = calyx, p = peduncle, α L and β L = lobes, cb = central body neuropile, pb = protocerebral bridge.

Right: Main acoustical reactions of male crickets due to stimulation in the protocerebrum (patterns are redrawings from oscillograms). a, b = inhibition of calling with post-inhibitory activation (1); c = initiation of calling with post-excitatory depression (2); d, e = aggressive sounds with regular and irregular arranged trills; f, g, h = different examples of atypical sounds with reappearance of rhythmicity in the post-stimulatory phase of the continuous song (3).

frequently and sang at last in a rhythm comparable to calling, as far as the number of pulses in one trill, the pulse intervals and the trill intervals are concerned. Atypical songs demonstrate firstly that the mesothoracic motor centre does respond to imposed atypical cerebral patterns of excitation, and secondly, that there is no 1 : 1 relation between the pattern of shocks and that of wing movements. The reappearance of 4 or 5 pulsed trills in the post-stimulatory phase indicates again a temporary distortion of pattern formation which takes place within or close to the central body neuropile.

III. Summary

Electrical stimulation applied to different areas of the cricket brain has revealed structures within the brain which are part of a neural mechanism responsible for inhibitory and excitatory control of sound production and associated behaviour. Some of the neurones seem to be essential in the formation of the song patterns in male crickets.

REFERENCES

Huber, F. (1955). Sitz und Bedeutung nervöser Zentren für Instinkthandlungen beim Männchen von *Gryllus campestris* L. *Z. Tierpsychol.* **12**, 12–18.

Huber, F. (1960). Untersuchungen über die Funktion des Zentralnervensystems und insbesondere des Gehirnes bei der Fortbewegung und der Lauterzeugung der Grillen. *Z. vergl. Physiol.* **44**, 60–132.

Huber, F. (1962). Central nervous control of sound production in crickets and some speculations on its evolution. *Evolution* **16**, 429–442.

Huber, F. (1964). The role of the central nervous system in Orthoptera during the co-ordination and control of stridulation. *In* "Acoustic behaviour of animals" (R. G. Busnel, ed.), pp. 440–488. Elsevier, Amsterdam.

Roeder, K. D. (1948). Organization of the ascending giant fibre system in the cockroach (*Periplaneta americana*). *J. exp. Zool.* **108**, 243–262.

Roeder, K. D. (1959). A physiological approach to the relation between prey and predator. *Smithson. Misc. Coll.* **137**, 287–306.

Epilogue

K. D. Roeder

Department of Biology, Tufts University, Medford, Massachusetts, U.S.A.

Perhaps the most striking impression left by this series of papers on the insect central nervous system is the wide variety of levels at which it is now being studied. Most of the modern tools and techniques of physiology such as electrophysiology, chromatography, flame photometry and electron microscopy have played a part in obtaining the results discussed today, and practically all the work was conceived during the past ten to fifteen years. In viewing our general progress it seems worth speculating about the extent to which its direction has been influenced by the availability of these techniques. I am incurably fascinated by gadgets, but I am also aware of their mute and sometimes baleful influence on what I do.

Until about thirty years ago physiological studies of the insect nervous system followed a different and more stereotyped pattern. Traces of this approach are detectable only in three out of the fifteen papers presented today. In this early work information about the functions of ganglia or parts thereof in controlling locomotion, reflexes and orientation was obtained by surgically removing these parts and recording changes in the behaviour of the deficient insect. Some of the names associated with this subtractive or difference method are Bethe, Kopec, von Buddenbrock, von Uexküll, Baldi, Baldus, and ten Cate, who also reviewed the main contributions in 1930. I owe my first acquaintance with the insect nervous system to this method. The dispersed and metameric organization of at least the more primitive orders made insects excellent subjects for such surgery, and more was learned about the functions of gross regions of the central nervous system in arthropods than in any other major group of animals. Furthermore, important generalizations about the central interaction of excitation and inhibition, and about input-output relations in orientation, were both explicit and implicit in these venerable studies, although the presence of electrical impulses in insect nerves had not been directly demonstrated.

On the debit side these early studies left us with some biases that are hard to live down. Conclusions about the function of a part were based on normal behaviour minus the behaviour shown after the part had been removed. This endowed information about the part with a sort of negative character which could be conceptualized only by thinking of the part as a *centre* controlling the behavioural difference. The centre concept, as I shall call it, is still applicable in a general way to the major regions of the insect central nervous system, but when one attempts to apply it to the finer (cellular) grain of ganglia it turns out to be a "will-o'-the-wisp" that has lured us into a conceptual fog not yet dissipated by the assorted matrix, net and field theories proposed in recent years. Thus, valuable as the ablation work was, it left us with an *idée fixe* whose easy comprehensibility—one place, one function—as well as, to a certain degree, validity, makes it very hard to replace. Today it is practically impossible to avoid using the word "localization" in talking about central nervous function, even when one can see no evidence for a "locus" of the neural organization concerned with a particular behaviour pattern, and the confusion will continue until the centre concept is replaced or modified by another concept of comparable heuristic value. I think that the centre concept was the outcome in some degree of the ablation method, and that today we are just as likely to become conceptually straight-jacketed by our present experimental methods.

As far as research on insect central nervous systems is concerned, the present era began with the introduction of electrophysiological methods. Adrian in 1930 was the first to record spontaneous nerve impulses from the isolated abdominal nerve cords of caterpillars and beetles. Prosser and Roeder followed with similar studies on other arthropods. This approach is the antithesis of the ablation method. Rather than being concerned with what the animal does after removal of a part, it deals only with events taking place in the part removed or isolated. These events invariably turn out to be complex in their turn, and methods are refined to deal with smaller and smaller parts. The rest of the animal—the main concern of the earlier workers—is neglected for the study of its parts.

Adequate testimony to the power and penetration of this approach and its tools—microelectrodes, photometry, chromatography, microchemistry, and electron microscopy—is provided by the papers of Narahashi, Treherne, Ray and Smith. Obviously, insect nervous systems must be subjected to this sort of scrutiny because it leads to greater understanding of the basic nature and commonality of living things. However, it seems inevitable that larger and larger taxa lose their identity as these synthetic "sense organs" narrow one's field of vision from whole animal to cell membrane and organelle. In one biological sense this is reassuring, that is to say, it constitutes "understanding"; in another it is vaguely disconcerting. Although we know that the cell is genotypically defined, its phenotype becomes blurred and unrecognizable when scrutinized with the technical "eyes" now available. What of the rest of the animal that was discarded? Do the phenotypic differences between taxa reside only in cellular organization? But cells are not bricks. The two biological concepts of commonality and difference, neatly expressed in the four words of the title "The Origin of Species", deserve an equal emphasis, and this is not provided alone by the molecular approach. Unless the analysis of living matter down to its common elements is accompanied by synthesis, or at least by a continuing concern with the organization of the complex system that is an intact animal, I feel that we are likely to end up in a fog of confusion as great as that in which the centre concept has left us.

For these reasons it is reassuring to find the central part of our proceedings occupied by papers concerned with "organs" of the nervous system. Boistel and Gahery, Sanborn and Senff, Miller and Weevers deal with cells complexed into receptor and effector organs whose operations are related specifically to regulatory functions of the insect nervous system. Their approach begins in the middle of the biological organization and can lead either to finer detail or to the role of the organ in the whole. However, this approach has its own special frustrations, as I know from personal experience. Sometimes one feels like a mechanic who spends his life working on ignition systems without ever having the opportunity to enjoy the rarefied atmosphere of electromagnetic theory or the pleasures of a complete automobile. Nevertheless, such experiments form a link between the commonality and the difference, and therefore are most likely to provide the substance of our knowledge of the working of the insect nervous system.

The papers by Hughes, Rowell and Huber constitute a fresh approach to the problems that were the concern of classical studies before 1930. Their sights are directed towards the total organism and its uniqueness of operation, although the weapons they sight belong to modern neurophysiology. Given certain neuronal properties, how does the system of neurones work? Their main problem is the conceptual one mentioned above—that of finding another workable dimension of neurone topology to replace or at any rate to augment the centre concept. If a new way of regarding neuronal interplay is to emerge, I think that work on insects is likely to play an important part. The dispersed nature of their ganglia and receptor fields and the parsimonious distribution of nerve cells in many of their behavioural patterns should favour the search for ways of narrowing the gap between what nerve cells do and how animals behave.

Learning in invertebrates or parts thereof is taking the attention of increasing numbers of biologists. Here we are bedevilled by semantic uncertainty about the meaning of the term. Previous experience modifies the performance of most neurones or neurone systems. Commonly this takes the form of adaptation or decreasing sensitivity in the presence of a repeated stimulus, or it may involve the opposite facilitation. The time course of adaptation is usually measured in milliseconds, as is the consequent process of disadaptation when the stimulus is withheld. There are a number of ways in which a neuronal system can be rendered more permeable by previous activity. For instance, it might contain an inhibitory component that adapts more rapidly on stimulus repetition than the excitatory components. Until this inhibitory component disadapts the system will remain more permeable after the cessation of stimulation. The time constants of such effects are usually less than a second. How long must the time constant of such a system be for it to be considered as a learning process? The time courses for sensory adaptation and disadaptation are roughly similar or symmetrical. Do we speak of learning only when the time for acquisition is vastly shorter than the time for retention? If such a system is discovered as part of the neural complex of a single ganglion, can it be considered *per se* as evidence of learning ability in the animal in question? Horridge and Hoyle examine these and similar questions in connection with neurophysiological experiments on cockroach ganglia. They demonstrate the potential value of insects in providing relatively accessible polyneuronal analogues of learning systems, but there is much still to be done before such analogues can be associated with adaptively significant learning in intact organisms.

Finally, I would like to make a few comments on the general significance of our proceedings and on the directions that seem to be receiving the least attention.

Some will criticize this assembly of papers on the insect central nervous system as a very parochial affair. Why select only this class of animals for special consideration? First, I will admit to this parochial attitude for myself at any rate, and I can find no explanation other than that I like working on insects. I suspect that this is true for many of us here, and that we became imprinted at a tender age to an insect collection. One could justify our fixation by enlarging upon the size and importance of the class, or upon the suitability of insects for neurophysiological work. This has been amply demonstrated by the preceding papers, and in any case the arguments seem specious when used in this connection, for I suspect that most of us became hopelessly entrapped by our early experiences.

The main danger of making a parochial beginning is that one may become even more confined. These proceedings show clearly that this is not the case, for our interests have ranged from ion fluxes and ultra-structure to learning and behaviour. Yet there are some awkward gaps.

The first gap, in my opinion, is in plain old-fashioned microscopic neuroanatomy. I always look with envy at the wealth of detail on the topography of tracts and nuclei in textbooks of mammalian neuro-anatomy, and wonder why insect nervous systems with their simpler organization have not been similarly treated. There is, of course, the work of Powers on *Drosophila*, but nothing comparable has been attempted on the Orthoptera, Blattaria, or Odonata, although they provide most of our subjects. Perhaps this sort of work should have been done fifty years ago and it is now too late, but much of our physiology is very shaky for this lack of structural foundation. A great opportunity still awaits those who can scan serial sections with intelli-gence and imagination.

A second gap concerns insect behaviour. Collecting information on behaviour requires a level of patience and passivity equal to that required of the microanatomist, and the two fields are similar in other ways. If carried out for themselves alone both can become immensely significant when related to the matters we have discussed today. By "behaviour" I do not mean merely the activities of insects under con-trolled laboratory conditions, although this may be an essential step. The nervous system is presumed to have become what it is through natural selection. Therefore, it seems only fair that the ultimate test of its operation should be under conditions as close as possible to those which channelled its evolution. I admit that this is often more of an ultimate objective than a practical possibility; my own studies of the mechanisms of evasive behaviour in cockroaches and moths present me with the somewhat awesome task of recording nerve activity from central ganglia without first startling the subject and adapting out its evasive behaviour.

Students of insect sensory physiology and orientation have somehow managed to establish a closer correlation between sense cell function and behaviour. This is particularly evident in the studies of vision and chemoreception. The same is true about cybernetic studies, but this field is limited by definition to behavioural criteria. Perhaps their greater success in connecting with normal behaviour stems from the fact that these fields deal primarily with the surface of their subjects, while experimental intervention in the central nervous system requires more radical disturbance of the body organization. Yet in spite of these difficulties, I feel that neurophysiology, even when it embraces only single nerve cells, loses much of its biological importance if the experimenter's horizon does not also include the whole animal in its natural context.

Author Index

Numbers in italics refer to pages on which references are listed at the end of the chapter

Subject Index

O

R